Collins *gem*

Ancient Greece

D0993111

David Pickering

First published in 2007 by
Collins, an imprint of
HarperCollins Publishers
77-85 Fulham Palace Road
London, W6 8JB

www.collins.co.uk

Reprint 10 9 8 7 6 5 4 3 2

All photographic images supplied by Corbis
Illustrations and artwork supplied by Cara Wilson

A catalogue record for this book is available from the British Library

Created by: SP Creative Design
Editor: Heather Thomas
Designer: Rolando Ugolini

ISBN: 978 – 0 – 00 – 723165 – 2

Printed and bound in China by South China Printing Co. Ltd

CONTENTS

INTRODUCTION   6

PART ONE: THE LAND OF THE GREEKS   9

The Greek world   10
Greek city-states   12

PART TWO: HISTORY OF ANCIENT GREECE   30

The Bronze Age   32
The Dark Ages   38
The Archaic Period   39
The Classical Period   42
The Hellenistic Period   46

PART THREE: RELIGION AND MYTHOLOGY   48

Gods and goddesses   48
Zeus   50
Hera   52
Aphrodite   53
Apollo   55
Ares   57
Artemis   59
Athena   61
Demeter   63
Dionysus   64
Hephaestos   66

Hermes   68
Hestia   70
Poseidon   71
Lesser gods   73
Companions to the gods   79
The Underworld   83
Mythical beasts   85
Heroes   90
Temples   99
Oracles   101
Cults and rituals   103
Festivals   105

PART FOUR: LIFE IN ANCIENT GREECE   108

Greek society   108
Government   111
Athenian politicians   114
Working life   118
Travel and transport   123
Family life   125
Education   130
Houses   132
Food and drink   136
Clothing and hairstyles   138
Make-up and jewellery   140
Leisure activities   142
Sport   144
Death in ancient Greece   148

PART FIVE: GREEK CULTURE   154

Culture and learning   154
Architecture   156
Sculpture   160

Pottery	162
Metalwork	164
Language	166
Literature	168
Theatre	173
Music and dance	182
Philosophy	184
History	191
Science and mathematics	193
Medicine	202

PART SIX: WAR IN ANCIENT GREECE	**208**

Armies	208
Weapons and armour	210
Tactics	212
Navies	214
Major wars	216
Alexander the Great	221

PART SEVEN: GREECE TODAY	**228**

Discovering ancient Greece	228
Ancient Greek archaeology	230
What you can see today	232
The cultural legacy	236
Museums	241

FIND OUT MORE	**246**

GLOSSARY	**249**

INDEX	**253**

INTRODUCTION

At first glance, the civilization of ancient Greece, with its ruined temples, vanished gods and age-old mythology, may seem to belong entirely to the past. Over the centuries, however, the culture forged by the ancient Greeks has had a profound influence upon the development of modern Western civilization, and in many respects their ideas and achievements remain highly relevant in the modern world.

THE INFLUENCE OF ANCIENT GREECE

The ideas of ancient Greek thinkers contributed greatly to the foundation of contemporary philosophy, politics, mathematics, science, architecture, literature and a host of other subjects. Even our modern languages contain thousands of words of Greek origin.

The gods and heroes of Greek mythology, meanwhile, have long been absorbed into the modern literary tradition, and continue to exert a powerful hold on the modern imagination, providing inspiration and material for a plethora of novels, plays and films.

The same may also be said of a small number of real people, such as Alexander the Great, and of the actual historical events of the period, encompassing a wide panorama of history, from the invention of democracy to the many wars that were waged both between Greek city-states and with enemies from beyond the borders of Greece.

THE CRADLE OF CIVILIZATION

Often described as the cradle of modern European civilization, ancient Greece continues to fascinate everyone, from archaeologists to tourists. We know a surprising amount about ancient Greece and the lives of its varied inhabitants, largely because theirs was one of the first civilizations to leave behind a detailed written and artistic record.

This book aims to summarize what we understand today about the world of the ancient Greeks, including their history, their religion and mythology, their daily lives, their culture and the wars they fought.

The Greek revival

Interest in the world of the ancient Greeks intensified in the late eighteenth and early nineteenth centuries, when it became fashionable for wealthy aristocrats and academics to visit sites in the classical world. Greek ideas were soon taken up with enthusiasm throughout the arts and sciences.

PART ONE

The land of the Greeks

The world of the ancient Greeks extended much further than mainland Greece, with the establishment of Greek colonies throughout the Mediterranean region and the Black Sea. Greece itself was not a single united country, but a host of independent city-states that shared a common culture.

The most powerful city-state
Athens, dominated by the temple of the Acropolis, was the most important of all the Greek city-states.

THE GREEK WORLD

Situated in the central Mediterranean, with easy access by sea to Egypt and north Africa, Asia Minor and what is now modern western Europe, Greece was positioned ideally as a centre of trade and cultural exchange. Its mountainous landscape, meanwhile, made invasion by outsiders difficult and allowed the various small city-states to develop in relative peace and largely independent of their neighbours.

COLONIZATION

Hot, dry and mountainous, mainland Greece offered little good farmland, so from an early date the Greeks explored far afield in search of places where they could grow crops to feed the population. Colonization of coastal areas throughout the Mediterranean and Black Sea region increased from around 750 BC, spreading Greek influence the length and breadth of the known world. By 600 BC, there were some 1,500 Greek colonies around the Mediterranean.

Thriving colonies were also established to the north and east, in what is now modern Bulgaria and Turkey, as well as on numerous islands in the Aegean Sea. There was also a strong Greek presence in north Africa, southern Italy, Sicily and southern France.

GREEK CULTURE

Everywhere the Greeks went, they took their culture with them, building temples, establishing trade links and communicating their ideas about religion, art, science and philosophy.

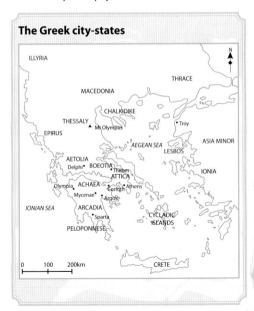

The Greek city-states

The Greek World

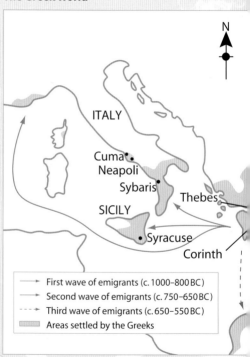

N

ITALY

Cuma
Neapoli
Sybaris

Thebes

SICILY

Syracuse

Corinth

→ First wave of emigrants (c. 1000–800 BC)
→ Second wave of emigrants (c. 750–650 BC)
⇢ Third wave of emigrants (c. 650–550 BC)
▨ Areas settled by the Greeks

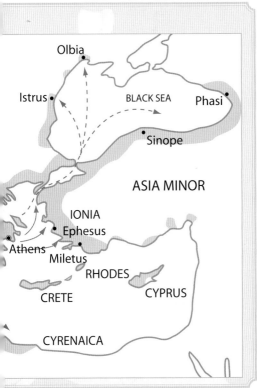

Olbia

Istrus

BLACK SEA

Phasi

Sinope

ASIA MINOR

IONIA

Ephesus

Athens

Miletus

RHODES

CYPRUS

CRETE

CYRENAICA

GREEK CITY-STATES

Greece was not one united country but a collection of independent city-states, all of which had their own patron gods, laws and constitutions. The first communities grew up where there was a good water supply and suitable farmland. These small communities later combined to form the first city-states, which typically included a city (usually a seaport) and the surrounding land with its villages and farms.

FIERCE INDEPENDENCE

Each city-state (or *polis* – hence the word 'politics') developed its own distinctive character. Cities were usually separated from their neighbours by mountains, valleys or the sea, which limited contact between them. However, city-states traded enthusiastically with each other and shared the same language, religion and other cultural features. Citizens were fiercely loyal to their particular state and, if called upon, would fight

Populations

Aristotle believed that city-states had to have more than ten citizens but less than 100,000, as citizens needed to know the people they voted for.

to defend their homeland. If attacked, citizens retreated to the acropolis (a citadel or walled area at the highest point of the city). The development of Greek civilization was regularly interrupted by squabbles between neighbouring states, although on rare occasions they allied to resist invasion by foreign armies (namely, against the Trojans in the late twelfth century BC, the Persians in the fifth and sixth centuries BC and Alexander the Great in the fourth century BC).

ATHENS

Athens was the largest and most powerful of the ancient city-states. At its peak in the Classical Period, it had over 250,000 inhabitants and was a centre of culture and learning. Named after its patron, the goddess Athena, it developed around a hill called the Acropolis, which was originally a fort but later became the religious heart of the state and the site of the Parthenon and other temples.

Athenian culture

The Athenians were responsible for many of the great advances of ancient Greek civilization, from democracy to philosophy and the theatre. It was home to celebrated philosophers, politicians and artists, and was a thriving trading centre, with business being conducted in its marketplace (the *agora*). The funds provided by such trade paid for the construction of many great buildings and other treasures for which Athens became famous.

Classical Athens

This map shows the layout of classical Athens with the Acropolis and Parthenon.

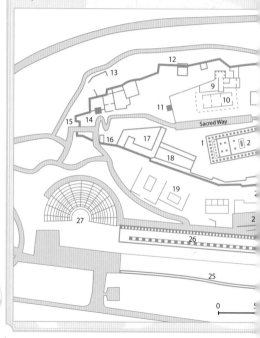

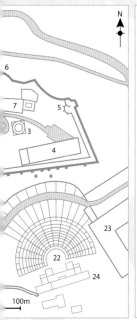

N

1 Parthenon
2 Statue of Athena Parthenos
3 Altar of Rome and Augustus
4 Acropolis Museum
5 Belvedere
6 Sanctuary of Eros
   and Aphrodite
7 Boukoleion
8 Altar of Athena Polias
9 Erechtheion
10 Porch of Caryatids
11 Statue of Athena
   Promachos
12 House of Arrhephoroi
13 Klepsydra (sacred cave)
14 Monument of Agrippa
15 Beule Gate
16 Temple of Athena Nike
17 Precinct of Artemis
   Brauronia
18 Chalkotheke
19 Temple of Themis
20 Sacred Spring
21 Incubation area
22 Theatre of Dionysus
   Eleuthereus
23 Odeum of Pericles
24 Sacred Precinct of
   Dionysus Eleuthereus
25 Aqueduct of Peisistratos
26 Stoa of Eumenes
27 Odeum of Herodes Atticus

100m

Defence

The Athenians possessed a large fleet and forged alliances with other states around the Aegean Sea. Their army was less strong, however, and was run on a part-time basis, with ordinary citizens being expected to help defend the state when it was attacked.

The temple ruins at Corinth, which was second only to Athens among the Greek city-states.

Ancient ruins

The ruins of some of the most important buildings of ancient Athens can still be seen in the heart of the modern city. These include the buildings on the Acropolis and the restored *agora*, which includes the Theseum.

CORINTH

The city-state of Corinth was situated in southern Greece, in the Peloponnese region on the Isthmus of Corinth. It was first settled some time before 3000 BC and emerged as a thriving centre for trade during the eighth century BC. In due course, it became the second largest and richest of the Greek city-states, after Athens. Rivalry between Corinth and Athens eventually led to the Peloponnesian War. Corinth was destroyed by the Romans in 146 BC.

CRETE

Crete, the largest of all the Greek islands, was an independent territory in the eastern Mediterranean. It was colonized around the sixth millennium BC and became home to the powerful Minoan civilization. Many Minoan relics have been found on the island, notably the royal palace at Knossos. Crete traded with many other Greek states but otherwise kept out of their quarrels and had little impact on later Greek civilization, although Cretan pirates were much feared. The island fell to Rome in 67 BC.

EPHESUS

Ephesus was a Greek city-state in Ionia, on the west coast of Asia Minor. An important trading centre, it thrived under the Persians, against whom it had waged war previously, and subsequently under the Romans. Its glories included a huge temple that was dedicated to the goddess Artemis. It was sacked by the Goths in 262 AD.

MILETUS

Miletus was an Ionian city-state that was founded around 1000 BC. Its early wealth was based on the trade in wool. Its enterprising merchants forged many links throughout the region and promoted Greek

Atlantis

Ancient Greek legend told of a lost island civilization west of Gibraltar. Said to have ruled over much of Europe and Africa, it was either destroyed by an alliance of free cities or else by an earthquake, and vanished forever below the waves. It is thought that the Greeks may have inherited the myth from the ancient Egyptians.

colonization of the Black Sea area. Having produced many leading thinkers of the sixth century BC, it declined after being sacked by the Persians in 494 BC.

RHODES

The island state of Rhodes was colonized by Dorian Greeks before 1000 BC. It became independent of Macedonia around 408 BC and enjoyed its peak in the third century BC. After successfully resisting a siege in 305–304 BC, the citizens of Rhodes celebrated by building the Colossus of Rhodes, a huge 31-metre-high statue of the sun god Helios bestriding the entrance to the harbour. One of the Seven Wonders of the World, it was destroyed by an earthquake in 244 BC.

SPARTA

The city-state of Sparta was the capital of Laconia in southern Greece. It grew out of Dorian settlements in the tenth century BC and by 700 BC dominated much of Laconia and Messenia. It was ruled by two hereditary kings, who were assisted by magistrates (the *ephors*) and a council of elders (the *gerousia*), as well as an assembly of citizens (the *apella*).

Overleaf: The Spartans valued military prowess above all other qualities, and their army was the most powerful of all those defending the Greek city-states.

A strict military state, Sparta boasted ancient Greece's strongest army. All males were brought up to serve in the army, young boys being trained as soldiers from the age of seven. Babies thought too weak to become soldiers were left to die in the open. The Spartans clashed frequently with the Athenians in the fifth century BC, finally defeating them in the Peloponnesian War (431–404 BC). In 371 BC, however, the Spartans themselves were defeated by the Thebans at Leuctra, triggering the decline of their influence. The Spartans placed little value on education or learning, and thus they had limited impact upon Greek culture. Their city was eventually destroyed by the Visigoths in 396 AD.

SYRACUSE

The seaport of Syracuse was the focus of Greek civilization in Sicily. Founded by Greeks from Corinth in 734 BC, it became an important cultural centre. Notable people born there included the poet Theocritus and the mathematician and scientist Archimedes. Syracuse fell to the Romans in 212 BC after a three-year siege.

THEBES

Situated in Boeotia in central Greece, the city of Thebes was founded in Mycenaean times and was celebrated as the setting of many legends in Greek drama. It was the most powerful city-state in the region until 480 BC, when it became notorious for siding with the Persian invaders against neighbouring states. It regained its dominant position in 371 BC after its army defeated the Spartans at Leuctra. The Thebans opposed Macedonian expansion and their city was destroyed by Alexander the Great in 336 BC.

THE PELOPONNESIAN LEAGUE

The various states that occupied the southern part of mainland Greece – the peninsula known as the Peloponnese, which was joined to the rest of Greece only by the Isthmus of Corinth – eventually realized that they would stand a much better chance of fending off invasion by the Athenians or other enemies if they formed alliances with each other.

Even the Spartans of Laconia, whose army was the strongest in the region, saw the sense in uniting with their neighbours to defend their homeland. Accordingly, Sparta became the senior partner in an alliance of states known as the Peloponnesian League. The other members included Elis, Arcadia, Corinth and Megara. These states were allowed to maintain their independence from Sparta, as long as they provided military assistance when requested to do so. It was this grouping of states that, after many battles, finally defeated the Athenians during the Peloponnesian War of 431–404 BC.

TOWNS AND CITIES

The majority of the towns and cities of ancient Greece were built very close to the sea, which provided a convenient means of access to the rest of the world. Most settlements were sited on a hill, which made them easier to defend from attack. Many of the most important structures, such as temples and buildings of religious or civic significance, were grouped on the crown of this hill, which was called the *acropolis* (meaning 'high city'). Further protection was provided by a high wall, which encircled both the acropolis and the surrounding built-up area.

COMMON FEATURES

Certain features were common to most Greek towns and cities. These included one or more temples, an *agora* (the site of the main marketplace and craftsmen's shops), a theatre, a gymnasium and a prison, as well as numerous private houses of various types and sizes.

Greek towns and cities were busy places, with people from the surrounding countryside and further afield mixing with locals to trade goods, buy slaves and fulfil their various obligations to the state. The streets, which were narrow and often steep, were usually crowded with people going about their daily business.

GREAT CITIES

Athens is often thought of as the centre of ancient Greek civilization, but there were many other notable cities through the Greek world that could boast a similar reputation and wealth. Elsewhere in mainland Greece, for instance, the city of Thebes (in Boeotia) was a centre of the ancient Mycenaean culture, while Corinth, with its large acropolis and harbour, could rival Athens as a cultural and commercial hub. Epidaurus (in the northeastern Peloponnese) was remarkable for its great buildings erected in the fourth century BC, when it was the focus of the cult of Asclepius, the god of healing.

The jewel in the crown of the later Hellenistic world was undoubtedly Alexandria, the city that was founded in northern Egypt by Alexander the Great in 332 BC. It was a centre of European culture long after the fortunes of the ancient Greeks had been eclipsed by the rise of the Romans.

Town planning

With so many people living in close proximity, it was fortunate that the Greeks had a sophisticated knowledge of town planning. Excavations of ancient Greek settlements have uncovered extensive drainage systems that ensured relatively healthy living conditions.

A typical Greek town

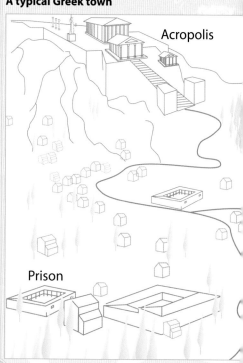

Acropolis

Prison

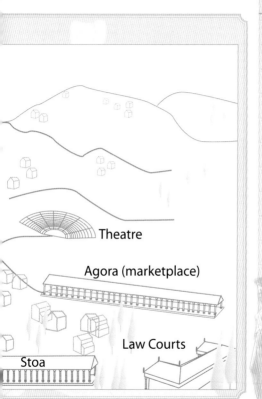

Theatre

Agora (marketplace)

Law Courts

Stoa

PART TWO

History of ancient Greece

Ancient Greek civilization had its roots in the Minoan and Mycenaean cultures that emerged during the Bronze Age, around 4,000 years ago. The golden age of ancient Greece followed during the Classical Period, which spanned the fourth and fifth centuries BC, and it ultimately ended with domination by the Macedonians and then the Romans in the first century BC.

Minoan wall painting

This Minoan wall painting known as 'The Blue Ladies' is a tantalizing relic of early Greek culture, dating from around 1600 BC.

THE BRONZE AGE

c. 2900–c. 1100 BC

c. 2900 BC   As the population of the region increases, villages become towns and craftsmen learn to make objects of bronze.

c. 2500 BC   The city of Troy is founded.

c. 2000 BC   The first Greek-speaking people arrive on mainland Greece.

c. 1930 BC   The first palaces are built by the Minoans on the island of Crete, thus marking the emergence of the earliest Greek civilization.

c. 1700 BC   The Minoan palaces are destroyed in an earthquake but are then rebuilt.

c. 1600 BC   The Mycenaean culture emerges on the mainland of Greece.

c. 1500 BC   A volcanic eruption on the island of Thera causes widespread damage.

c. 1450 BC   Many Minoans on Crete are killed in a second eruption. The capital Knossos is burned down soon afterwards and is never rebuilt.

c. 1400 BC   Crete is plundered by the Mycenaeans.

c. 1250 BC   The Trojan War between the Greeks and the Trojans begins.

c. 1200 BC   Mycenaean power declines and the Mycenaean cities are slowly abandoned.

THE ORIGINS OF ANCIENT GREECE

The eastern Mediterranean was first settled by nomadic hunter-gatherers around 40,000 BC. By 6300 BC they had learned how to farm the land and were able to make simple pots and other objects using stone tools. It was not until they learned how to use bronze to make weapons and other useful items, however, that the first significant steps were made towards building a civilization that was the equal of those already prospering in Egypt, Sumeria and elsewhere.

At around the same time, small isolated communities began to forge links and pave the way for the powerful city-states. Among the significant innovations of the period was the introduction of the first sailing vessels, which promoted trade and cultural communication.

EARLY GREEK CULTURES

The earliest Greek civilizations were those of the Minoans on the island of Crete and the Mycenaeans on the mainland of Greece. It is evident from the remarkable archaeological remains of the palaces and the other major structures built by these peoples that they achieved impressive levels of sophistication. Their artistic, religious and cultural ideas were to have a profound effect upon the later development of Greek civilization.

THE MINOANS

The Bronze Age Minoan civilization on the island of Crete is best known from the ruins of luxurious palaces at Knossos, Zakro, Mallia and Phaestos. Their grand architectural designs and wall paintings are proof of the wealth of the Minoans (achieved chiefly through trade) and also of their artistic achievements. Their craftsmen included highly skilled architects, potters, painters, stonemasons, goldsmiths and jewellers. They also had knowledge of hieroglyphics and other scripts. Among the most important deities was a snake goddess, although it also appears that bulls were sacred.

The Minoan civilization never recovered from the catastrophic volcanic eruption on the nearby island of Thera (now called Santorini) in 1450 BC. Fallout from the volcano wrecked agriculture on Crete and left the region open to increasing Mycenaean influence.

The land of King Minos

The Minoan civilization was named (in the early twentieth century) after the legendary King Minos, who is supposed to have kept the fearsome bull-headed Minotaur in a labyrinth under his palace. The excavated palace at Knossos has been identified as his (though 'Minos' may simply have been a royal title).

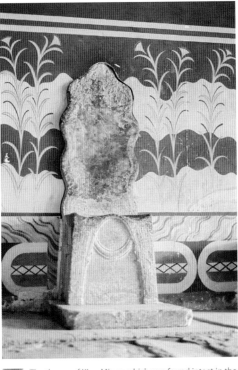

The throne of King Minos, which was found intact in the royal palace at Knossos in 1894.

THE MYCENAEANS

The first Greek civilization on the mainland was that of the Mycenaeans, a warrior people who flourished between 1600 BC and 1100 BC. After the collapse of the Minoan civilization, the influence of the Mycenaeans spread through the eastern Mediterranean area. They became very wealthy and built richly decorated palaces and other buildings, of which only fragments – such as the famous Lion Gate marking the entrance to the citadel of the city of Mycenae – have survived. The impressive size of the walls surrounding Mycenaean settlements led later Greeks to think they had been built by giants. Other features of Mycenaean culture included wonderful wall paintings and metal goods.

Mycenaean culture

Although the Mycenaeans lived in numerous small kingdoms, they shared the same language and similar religious beliefs. The dead were buried with precious objects in a shaft grave or in a beehive-shaped tomb (called a *tholos*). Poor harvests and repeated invasion

United against Troy

Around 1250 BC ,the early Greek city-states united under the Mycenaean king Agamemnon to fight Troy in the Trojan War.

by neighbouring peoples led to a gradual decline of Mycenaean culture around 1200 BC.

The Lion Gate at Mycenae is one of the more substantial remnants of Mycenaean culture.

THE DARK AGES

c. 1100–800 BC

c. 1050 BC   The first Greek settlements are founded on the west coast of Asia Minor.
c. 900 BC   The city-state of Sparta is founded.
c. 850 BC   Greek communities throughout the region develop, each with their own distinctive culture.

After the decline of the Minoan and Mycenaean civilizations, ancient Greece entered a period of some 300 years of which relatively little is known, as the art of writing had been lost. Myths and details of the region's earlier history were probably passed on orally. Trading links were broken, standards fell in pottery and jewellery making, and few stone buildings were erected. The dead were buried with very simple objects, in contrast to the wonderful treasures found in earlier graves.

EMIGRATION TO IONIA

During this period there might have been a sizeable reduction in the population, perhaps due to famine, with many people venturing to the west coast of Asia Minor in search of a better place to settle. These colonies, in what became known as Ionia, were to become thriving centres of Greek trade and culture later on.

THE ARCHAIC PERIOD

c. 800–c. 500 BC

c. 800 BC   The first independent city-states are formed. The Greeks develop their own alphabet, based on that of the Phoenicians. Homer writes *The Iliad* and *The Odyssey*.

776 BC   The first Olympic Games are held at Olympia.

c. 750 BC   The Greeks found colonies around the Mediterranean and Black Sea.

c. 735 BC   The first Greek colonies are established in Sicily.

c. 730 BC   The Spartans expand their territory in the Peloponnese region between 740–720 BC.

c. 700 BC   The epic poet Hesiod is born.

c. 650 BC   The first of a series of tyrant kings seizes power in Greece. The first coins are introduced.

c. 630 BC   The Spartans put down a rebellion by the Messenians c. 630–613 BC.

621 BC   Draco introduces strict new laws in Athens.

c. 594 BC   Solon reforms the Athenian political system.

c. 590 BC   Development of the Ionic and Doric orders of architecture.

585 BC	The philosopher and mathematician Thales predicts a solar eclipse.
c. 584 BC	The mathematician Pythagoras is born.
547 BC	The Persians seize Ionia.
540 BC	Anaximander, author of the first philosophical treatise and maker of the first maps, dies.
c. 530 BC	The Persian Wars begin.
525 BC	The playwright Aeschylus is born.
508 BC	The rule of the tyrant kings of Athens ends and democracy is introduced.

TRADE LINKS AND INFLUENCE

The fortunes of ancient Greece revived considerably around 800 BC with the reforging of international trade links with Egypt, Syria, Carthage and Tyre. This period also witnessed the rediscovery of writing and the spread of Greek influence throughout the known world.

CITY-STATES

Towns grew into city-states, with their own laws and government. The increasing wealth of such communities funded the development of craft and scholarship. During the Archaic Period, the Greeks began to make major steps forward in philosophy, science and the arts. Another notable landmark was the introduction of the world's first democracy in Athens in the sixth century.

COLONIZATION

As the population continued to grow and mainland Greece struggled to feed its inhabitants and became increasingly overcrowded, many citizens ventured overseas to settle in the new Greek colonies that were springing up throughout the Mediterranean region and around the Black Sea.

Squabbles between the city-states also encouraged much of the population to go abroad, taking their language, culture and scientific ideas with them. Many of the colonies became extremely wealthy and produced some of ancient Greece's greatest thinkers.

FOUNDATIONS OF GREATNESS

The Archaic Period saw the foundations laid for future greatness, with fundamental changes in Greek society and the establishment of the city-states. The first ones appeared in central Greece, the eastern Peloponnese and on the western coast of what is now Turkey (then known as Ionia). These were to remain the most important sites of Greek culture for 500 years. The period ended with Sparta, as head of the alliance known as the Peloponnesian League, being widely acknowledged as the foremost city-state, although the Athenians were already emerging as a leading rival and were soon to enter a Golden Age.

THE CLASSICAL PERIOD

c. 500–336 BC

499–494 BC	The Greek colonies in Ionia rebel against the Persians.
490 BC	The Persians invade the Greek mainland but are beaten at the Battle of Marathon.
480 BC	The Persians invade once more and beat the Greeks at Thermopylae. Athens is sacked, but the Persian fleet is then destroyed at the Battle of Salamis.
479 BC	The Persians evacuate Greece after defeat by the Athenians at the Battle of Plataea.
479–431 BC	Athens enjoys its Golden Age.
478 BC	The Delian League of Greek states is formed to resist Persia.
469 BC	The philosopher Socrates is born.
461–446 BC	The First Peloponnesian War is fought between Athens and Sparta. The Athenians build walls to protect from attack their link with the port of Piraeus.
460 BC	The physician Hippocrates is born.
449 BC	The Delian League of Greek states makes peace with Persia.
447–432 BC	The Parthenon is built on the Acropolis in Athens.
443–429 BC	Pericles governs as head of state in Athens.

c. 440 BC	Herodotus writes his histories.
431 BC	The Second Peloponnesian War breaks out between Athens and Sparta.
430 BC	Athens is hit by plague.
429 BC	The statesman Pericles dies.
415–413 BC	An Athenian expedition against Syracuse results in a disastrous defeat.
406 BC	Death of the playwrights Sophocles and Euripides.
404 BC	The Peloponnesian War ends in Spartan victory.
403 BC	The Spartans allow the restoration of democracy in Athens.
400 BC	Spartan power is at its greatest.
c. 400 BC	Plato writes on government.
399 BC	War breaks out between Sparta and Persia. Socrates is tried and forced to commit suicide.
395–387 BC	The Corinthian War is fought between Sparta and the allied armies of Corinth, Athens, Argos and Thebes.
386 BC	Plato founds the Academy in Athens.
384 BC	The philosopher Aristotle is born.
371 BC	The Thebans defeat the Spartans at Leuctra.
359 BC	Philip II is crowned king of Macedon.
356 BC	Alexander the Great is born.
347 BC	The philosopher Plato dies. Alexander the Great studies under Aristotle.

338 BC	Greece loses independence after defeat in battle against Macedon.
336 BC	Alexander the Great becomes king of Macedon.
334 BC	Alexander the Great begins eleven years of campaigning in Asia Minor.
323 BC	Alexander the Great dies at Babylon.

During the Classical Period, trade in the region flourished, bringing vast wealth to the city-states. Once the threat posed by the Persian Empire had been successfully fended off, ancient Greece became the leading power in the known world. Athens, the home of the world's first democracy, attracted the finest artists, philosophers, writers and scientists and was justly celebrated for its artistic and cultural brilliance.

THE GOLDEN AGE

Ancient Greek civilization reached a peak in the years following the defeat of the Persians in 479 BC. Great achievements were made in the areas of literature, architecture, the theatre, science and philosophy. Among the great names that are associated with the period are the Athenian leader Pericles, the playwrights Sophocles and Euripides and the historian Herodotus. The ideas of the Greek philosophers, in particular, were to become the foundation blocks for future western European civilization.

ATHENIAN DECLINE

The outbreak of the Peloponnesian Wars between Athens and Sparta signalled an end to ancient Greece's Golden Age. Athens was struck by plague and following some further setbacks – such as the failure of a military expedition against Syracuse on Sicily – it underwent political upheaval, leaving it vulnerable to attack. It was besieged and fell to the Spartans. Although democracy was soon restored, Athens never regained its former glory, but Sparta fared little better. Within 50 years, Sparta and Athens, together with the rest of Greece, fell under the domination of the Macedonians.

THE PERSIAN EMPIRE

Before the Athenians could enjoy what was to become their Golden Age, they had to see off the threat posed by invading Persians. The Persians came originally from what is now Iran, and by 485 BC their empire had expanded to take in Assyria, Egypt and Lydia, including the Greek colonies of Ionia. The Persian army was formidable. Most feared of all were the Immortals, an elite regiment of 10,000 warriors. Persian invaders were defeated at Marathon in 490 BC but overcame heroic Greek resistance at Thermopylae in 480 BC, going on to destroy Athens. The Persian threat was finally fended off after the decisive Greek victories at sea at Salamis in 480 BC and on land at Plataea in 479 BC.

THE HELLENISTIC PERIOD

323–30 BC

323–322 BC The Greek states remain under Macedonian control after defeat in the Lamian Wars.

322 BC Death of Aristotle and Demosthenes.

287 BC The scientist Archimedes is born.

279 BC Northern Greece is invaded by the Gauls.

224 BC The Colossus of Rhodes is destroyed by an earthquake.

214–205 BC War breaks out between Macedon and Rome.

212 BC Rome conquers Syracuse and Sicily; Archimedes is killed by a Roman soldier.

197 BC The Macedonians are defeated by Rome and lose control of Greece.

168 BC Macedon is conquered by the Romans.

146 BC Corinth is conquered by Rome, and Greece becomes a province of the Roman Empire.

Greek rulers of Egypt

The Greek influence was particularly strong upon the later history of Egypt, which was governed by Alexander's general Ptolemy and his descendants, the last of whom, Cleopatra VII, died in 30 BC.

Greece remained under the control of Macedon for some 150 years after conquest by Philip of Macedon and the reign of Alexander the Great. The region was subsequently fought over by various rivals in the course of squabbles between Alexander's successors.

GREECE UNDER THE ROMANS

When the Romans replaced the Macedonians as the dominant power in the Mediterranean, Greece fell under the shadow of Rome (which had begun by conquering the Greek city-states in southern Italy). The Romans, however, admired Greek culture and copied many Greek ideas about architecture, literature, theatre, religion, philosophy and science, and in so doing they passed them on to the modern world.

 Alexander the Great's reputation as a military leader has endured through the centuries.

PART THREE

Religion and mythology

The ancient Greeks thought that their everyday lives were at the mercy of the whims of a host of gods and goddesses. Belief in the gods was central to ancient Greek culture, and keeping the gods happy through worship at the many temples, shrines and festivals was important. If shown due respect, the gods might then answer the requests of mortals at their oracles.

The Parthenon
The Parthenon temple in Athens remains a symbol of ancient Greek civilization.

GODS AND GODDESSES

The ancient Greeks believed in a host of gods and goddesses, each representing a different aspect of life or nature. Ordinary Greeks called on the gods for help in times of difficulty, such as childbirth, famine or death.

THE GODS' BEHAVIOUR

The gods often behaved like ordinary mortals and many tales were told of their jealousies, quarrels and love affairs. The gods were usually depicted in human form, although Greek mythology also included numerous demi-gods or fantastic beasts who could be part-human and part-animal in form. In their dealings both with each other and with mortals, the gods were fickle and could be cruel or kind. Some helped favoured

The gods and men

As far as the ancient Greeks were concerned, there was no barrier between the divine and mortal worlds. The gods were believed to take a direct part in the affairs of mortals, and to control all aspects of the natural world, from the passage of the seasons to the occurrence of earthquakes and volcanoes. It was felt to be appropriate to honour the gods with sacrifices of animals and crops.

Overthrow of the Titans

Cronos was the king of the Titans. To defy a prophecy that he would be overthrown by his own children he ate them at birth. His son Zeus, however, was rescued by his mother Rhea, who gave Cronos a stone to eat instead, and Zeus went on to fulfil the prophecy.

mortals to gain a place among the heroes of Greek mythology, while others punished those who offended them with great severity.

CREATION OF THE UNIVERSE

The Greeks believed that the Earth (called Gaia) was created out of Chaos and gave birth to Uranos (the sky), whom she married, and Pontus (the sea). The children of Gaia and Uranos included the Titans, a race of giants who ruled the Earth until they were overthrown by Zeus and the twelve Olympian gods (so-called because they were said to live on Mount Olympus).

Hera, Zeus, Hestia, Poseidon, Pluto and Demeter were children of Cronos and Rhea, while Aphrodite was a daughter of Uranos, and Ares a son of Hera. The rest of the Twelve Olympians (Artemis, Apollo, Athene, Hermes and Dionysus, who replaced Hestia as one of the Olympian gods when she resigned) were children of Zeus.

ZEUS

KING OF THE GODS

Father: Cronos. **Mother**: Rhea. **Brothers**: Hades, Poseidon. **Sisters**: Demeter, Hera, Hestia. **Wives**: Metis, Themis and Hera. **Lovers**: Numerous goddesses and mortals. **Children**: Artemis, Apollo, Athena, Hermes, Dionysus, Herakles, Helen and many others. **Symbols**: Thunderbolt, eagle, oak tree. **Roman name**: Jupiter.

Zeus was the most powerful of all the Greek gods and goddesses and was worshipped as the ruler of the earth and sky. He was said to have brought order to the world by defeating his father, Cronos, and the Titans, the race of giants who previously ruled the universe. From his court on Mount Olympus, Zeus presided over the affairs of gods and mortals alike, often intervening personally to defend his favourites and punish those who incurred his wrath.

Leda and the swan

Typical of Zeus's love affairs was his conquest of Leda, the faithful wife of King Tyndareus of Sparta. Zeus seduced the unsuspecting woman in the form of a swan: she later gave birth to two sets of twins, the girls Clytemnestra and Helen and the boys Castor and Pollux.

The statue of Zeus in the temple of Zeus at Olympia was one of the Seven Wonders of the World.

Zeus was depicted as a strong, powerfully-built middle-aged man with a beard. He was worshipped as a defender of Greek civilization, upholding such virtues as justice, law, friendship and hospitality. In addition to ruling the seasons and determining the courses of the stars through the heavens, he also controlled the weather and punished wrongdoers by hurling thunderbolts (made for him by the Cyclopes) at them.

HUMAN FAILINGS

Zeus could, however, be bad-tempered and he often fell prey to various human failings. He had many love affairs with both goddesses and beautiful mortal women, visiting them as a bull, a swan or as a shower of gold, among other disguises. These episodes caused much distress to his wife Hera.

HERA

QUEEN OF THE GODS

Father: Cronos. **Mother**: Rhea. **Brothers**: Zeus, Hades, Poseidon. **Sisters**: Demeter, Hestia. **Husband**: Zeus. **Children**: Hephaestos, Ares and others. **Symbols**: Peacock, pomegranate. **Roman name**: Juno.

As the third and most important wife of Zeus, Hera was respected as the queen of the gods. Beautiful and proud, she was worshipped as the goddess of women, marriage, motherhood and children. Her own marriage, however, was unhappy due to her husband's repeated love affairs with other women, both mortal and immortal. Hera showed her jealous anger at this behaviour by nagging her husband for leaving her on her own so often and by punishing many of his lovers, including Leto, Semele and Io, sometimes with considerable cruelty.

Deadly snakes

When Zeus fathered the hero Herakles by Alcmene, Hera showed her resentment by sending deadly snakes to kill the baby. The infant Herakles, however, managed to strangle them and again escaped unscathed when Hera later sent the hundred-headed Hydra against him.

APHRODITE

GODDESS OF LOVE

Father: Uranos or Zeus. **Husband**: Hephaestos.
Lovers: Ares, Hermes, Dionysus, Adonis and others.
Children: Eros, Priapus, Phobos, Aeneas and others.
Symbols: Roses, doves, sparrows, dolphins and rams.
Roman name: Venus.

Aphrodite was the Greek goddess of love and beauty and was commonly depicted as a beautiful young woman, usually naked or semi-naked. Legend had it that she was born out of the sea after some drops of the blood of Uranos, the god of the sky, fell from the heavens when he was killed by his son Cronos. Called Aphrodite because it meant 'born from the foam', she floated to dry land on a scallop shell, coming ashore in Cyprus or on the southern Greek island of Cythera. A less colourful version of her birth claimed that she resulted from the union of Zeus and Diana.

UNFAITHFUL

Charming, beautiful and irresistible to gods and mortals alike, Aphrodite was also unfaithful and the cause of much unhappiness. She betrayed her husband Hephaestos with various lovers, including the beautiful youth Adonis, and was not a good mother. She also

The girdle of Aphrodite

Aphrodite possessed a magic girdle (belt) with the power to make any mortal who wore it irresistibly desirable.

liked to meddle in the affairs of others and it was largely her interfering that led to the outbreak of the Trojan War. She often punished beautiful rivals who boasted of their good looks and on other occasions she ordered her son Eros to fire magic arrows at certain men or women with the result that they fell in love.

Aphrodite is depicted here with her lover Ares and her son Eros.

APOLLO

GOD OF LIGHT

Father: Zeus. **Mother**: Leto. **Brothers**: Hermes, Dionysus, Herakles and others. **Sisters**: Artemis, Athena, Helen and others. **Children**: Asclepius, Orpheus, Aristeus and Philammon. **Symbol**: Laurel tree. **Roman name**: Apollo.

As well as being the god of light, Apollo was the god of the Sun, truth, music, poetry, science, healing, arts and education. The twin brother of Artemis, he was depicted as a beautiful young man with curly hair and no beard, and he had numerous love affairs with both goddesses and mortals while still a reckless youth. Not all of these were successful, however: when he fell in love with the nymph Daphne, for instance, she ran away from him and called on Zeus to help her. Just as Apollo caught up with her, Daphne was transformed by Zeus into a laurel tree and thus escaped his lust.

GIFT OF PROPHECY

Apollo was often worshipped in his role as the god of the Sun and was sometimes called Phoebus ('shining'). Like the Sun, he was credited with both life-giving and destructive properties. He was also believed to have the gift of prophecy and as he grew older and calmer employed his powers of prediction and healing for

A god's vengeance

Apollo could be vengeful against women who rejected him. When Cassandra refused his advances, he gave her the gift of prophecy on the condition no-one would believe her.

good, becoming known as a protector of animals, especially sheep and goats. Worshippers could consult him through the oracle at Delphi, which he made his own after killing his mother's enemy – the serpent Python – when it was sheltering in the shrine. When not on Mount Olympus, Apollo was said to spend his time on Mount Parnassus, the home of the nine Muses.

Bernini's famous sculpture shows Apollo in pursuit of the nymph Daphne.

ARES

GOD OF WAR

Father: Zeus. **Mother**: Hera. **Brothers**: Hephaestos and others. **Sisters**: Eileithyia, Hebe. **Lovers**: Aphrodite and others. **Children**: Deimos, Eros, Phobos, Penthesilia and others. **Symbols**: Burning torch, dogs, spear, vultures. **Roman name**: Mars.

The god of war, rage and violence, Ares was a terrifying figure who inherited all the fury and resentment of his mother Hera. He was depicted as strong, fierce, unruly and short-tempered. However, his reputation for violence meant that he was not the most popular of the gods among ordinary Greeks, who, as a result, had relatively few stories about him. One legend claimed that he was tried for murder on a hill in Athens, which was later named Areopagus after him.

NOTORIOUS LOVERS

Ares was described as being ruggedly handsome, and he had a notorious love affair with the goddess Aphrodite. By her and other lovers, he became the father of several notable heroes.

Aphrodite's outraged husband Hephaestos responded by making a metal net so fine that it was invisible and

> ### War cry
>
> Such was the ferocity of Ares that his war cry alone was said to be frightening enough to kill any mortal who heard it.

then threw it over Ares and Aphrodite as they lay in bed, trapping them, before summoning all the other gods to witness their shame.

COMPANIONS OF ARES

Ares was only one of several gods and goddesses associated with war. He represented war's violence and unpredictability, while his half-sister Athena symbolized strategic warfare and another sister, Enyo, represented bloodshed. Among his companions were minor gods of war, including Kydoimos, who embodied the din of battle, and Alala (whose name furnished Ares with his battle cry). His sons included the monstrous Cycnus, who murdered travellers in order to build a temple from their bones, until he was himself killed by Herakles.

Ares was particularly revered by the warlike Spartans, who claimed him as their ancestor, believing that the first Spartans sprang up when the teeth of the water-dragon Cadmus, one of the offspring of Ares, were sown in the ground. The Spartans erected a statue of him in chains, to bind him to their city forever.

ARTEMIS

MOON GODDESS

Father: Zeus. **Mother**: Leto. **Brothers**: Apollo, Hermes, Dionysus, Herakles and others. **Sisters**: Athena, Helen and others. **Symbols**: Cypress trees, deer, dogs. **Roman name**: Diana.

The twin sister of Apollo, Artemis was the goddess of hunting, wild animals and the countryside as well as being goddess of the moon. She was famed for her skill as an archer, and on occasion used her silver arrows to kill mortals and spread plague and death.

Her followers (most of whom were women) revered the goddess's powers of healing and believed that she was a loyal protector of young girls and pregnant women. Brides and women in childbirth commonly prayed and made offerings to Artemis, who had never known the pain of giving birth.

CHASTE HUNTER

Artemis was often depicted with a bow and arrow or driving in a chariot pulled by stags. It was said that she loved hunting, sports, playing with wild creatures and bathing in secluded streams. She never married and surrounded herself with companions who were chaste like herself. In character she was said to be cold and remote and she did not hesitate to punish those who angered her. When Oeneus, King of Calydon, failed to make proper sacrifices at the goddess's temple he found his country being terrorized by the Calydonian Boar, a giant boar sent by Artemis to destroy crops and kill his people. Many died hunting the creature before it was eventually killed and its tusks presented to Athena.

DIVINE PUNISHMENT

It was dangerous to offend Artemis. When Actaeon accidentally saw Artemis bathing naked, she turned him into a stag and ordered her hounds to hunt him to death. Artemis also featured prominently in the legend of the wondrously handsome youth Adonis. When Aphrodite killed Hippolytus, a favourite of Artemis, the latter got her revenge by sending a wild boar to kill the young man, who had rashly boasted that his skill as a hunter was superior to that of the goddess herself.

A terrible sacrifice

Another to experience the wrath of Artemis was King Agamemnon of Mycenae, who once killed a sacred deer and similarly claimed to be a better hunter than the goddess. Artemis subsequently stilled the wind when Agamemnon's fleet set sail for Troy, and it was only after the king sacrificed his daughter Iphigenia to her that the goddess allowed him to continue.

ATHENA

GODDESS OF WISDOM

Father: Zeus. **Mother**: Metis. **Brothers**: Apollo, Hermes, Dionysus, Herakles and others. **Sisters**: Artemis, Helen and others. **Symbols**: Owl and olive tree. **Roman name**: Minerva.

Athena (or Athene) was the Greek goddess of wisdom and war. Legend had it that Zeus swallowed Athena's pregnant mother Metis because of a prophecy that if she produced a son he would overthrow his father. When Zeus then suffered a headache he ordered the blacksmith god Hephaestos to crack his head open, and Athena sprang out fully grown and armed.

PROTECTOR OF HEROES

Athena presided over the arts, literature, learning and philosophy, and was also the patron of craftsmen of

Athena's city

Athena was the patron goddess of Athens. The temple called the Parthenon (meaning 'virgin') was dedicated to her and contained a towering statue of the goddess covered with gold and ivory.

various kinds, including ship builders, carpenters and potters. Her gifts to mortals included the double flute called the *aulos* and the skill to grow olive trees. She was often depicted as a spear-carrying warrior and carried a shield (the *aegis*) decorated with the head of the gorgon Medusa, which turned to stone all those who saw it. She protected heroes such as Herakles, Perseus, Jason and Odysseus, whom she guided on his way home from Troy. In her role as the goddess of victory, she was also known as Athena Nike.

Athena, the patron goddess of Athens, is depicted here armed with a spear and shield.

DEMETER

CORN GODDESS

Father: Cronos. **Mother**: Rhea. **Brothers**: Zeus, Poseidon, Hades. **Sisters**: Hera, Hestia. **Children**: Persephone. **Symbols**: Sheaf of barley or wheat, ear of corn. **Roman name**: Ceres.

Demeter was the goddess of crops and soil fertility. Legend described how she deliberately neglected the crops in protest after her daughter Persephone was kidnapped and carried off to the Underworld by Hades. The threat of famine persuaded Zeus to intervene and to order Hades to allow Persephone back to the world above. It was a condition of her release, however, that Persephone would spend a third of each year in the Underworld, during which time winter ruled the earth. Her return each year was greeted with the return of spring and summer, symbolizing new life after death.

Secret rituals

Demeter was believed to have taught King Triptolemus how to plough and sow grain at Eleusis near Athens. Eleusis later became famous for the secret rituals performed there in the goddess's honour.

DIONYSUS

GOD OF WINE

Father: Zeus. **Mother**: Semele. **Brothers**: Apollo, Hermes, Herakles and others. **Sisters**: Artemis, Athena, Helen and others. **Wife**: Ariadne. **Symbol**: Thyrsus (staff). **Roman name**: Bacchus.

Dionysus was the Greek god of wine, revelry, fertility, dancing and the theatre. Depicted as a jolly, bearded fat man with a huge belly and ivy in his hair – though he might also appear in the form of a bull, goat or as a beautiful youth – he taught mortals how to make wine. Celebrations held in his honour were attended chiefly by women and were often frenzied, with drinking and dancing. These wild events became an important feature of religious festivals and theatre performances. Dionysus replaced Hestia as one of the twelve Olympians when she resigned her place at the court of Zeus.

Captured by pirates

Legend related how Dionysus was once captured by pirates. When the sea was turned to wine, ivy sprouted from the mast, a bear appeared on deck and Dionysus himself was transformed into a lion; the pirates leapt overboard in terror and became dolphins.

FOLLOWERS OF DIONYSUS

Carrying a thyrsus (magic wand) which was made of fennel stalks, Dionysus was said to roam the countryside accompanied by satyrs (half-men, half-goat) and female followers called maenads ('wild women'). Another faithful companion was a fat, drunk, old man called Silenus, who shared wise advice with Dionysus.

 This Roman mosaic depicts the god Dionysus in the guise of a handsome youth.

HEPHAESTOS

BLACKSMITH TO THE GODS

Father: Zeus. **Mother**: Hera. **Brothers**: Ares and others. **Sisters**: Eileithyia, Hebe. **Wife**: Aphrodite. **Children**: Erichthonius, Periphetes, Cabiri, Palaemonius. **Symbol**: Blacksmith's hammer. **Roman name**: Vulcan.

Hephaestos was the god of fire and metalworking and the patron of craftsmen. According to legend, he was born lame and so ugly that his mother threw him into the ocean to drown. He was rescued, however, by a nymph and later became famous for his craftsmanship, fashioning weapons and jewellery from rock.

He was brought to Mount Olympus and there he was acknowledged as the god of fire and craftsmanship. He made many objects for the other gods with his mighty

Pandora's box

Legend credited Hephaestos with making Pandora, the first mortal woman. Zeus gave Pandora a box but forbade her from opening it. After arriving on Earth, Pandora could not resist peeking to see what was inside the box. As she lifted the lid all the ills of the world escaped to trouble mankind. All that was left inside was hope.

blacksmith's hammer, among them armour that made the wearer invincible and servants of gold who worked in his forge beneath Mount Etna in Sicily. He made a golden throne for Zeus and a shield that raised thunder and storms when shaken. He was also the maker of the winged helmet and sandals of Hermes, the magic girdle of Aphrodite, the armour of Achilles, the chariot of the sun god Helios and the bow and weapons of Eros.

A WRONGED HUSBAND

After Hephaestos was thrown off Mount Olympus by Hera for his ugliness he fell for several days and nights before coming to earth. He lived for nine years in a cave by the sea, perfecting his skills. On his return to Olympus, he presented Hera with a golden throne that trapped her when she sat in it. It was only when he was promised the lovely Aphrodite as his wife that he finally released the goddess. Later, though, Aphrodite angered her husband by having affairs with other gods and mortals.

DEFORMITY

It has been suggested that the deformed appearance of Hephaestos, as described by various writers and artists, may have been inspired by the effects of arsenic poisoning, which was an occupational hazard of blacksmiths during the Bronze Age. Arsenic was added to bronze as a matter of course to make it harder.

HERMES

MESSENGER OF THE GODS

Father: Zeus. **Mother**: Maia. **Brothers**: Apollo, Dionysus, Herakles and others. **Sisters**: Artemis, Athena, Helen and others. **Lovers**: Aphrodite, Polymele and others. **Children**: Pan, Eudorus, Hermaphroditus, Myrtilus, Cephalus, Autolycus and others. **Symbols**: Winged hat and sandals, staff. **Roman name**: Mercury.

Hermes was the messenger of the gods and was often sent on errands between Heaven and Earth. The patron of travellers and merchants, he was usually depicted wearing a traveller's hat and winged sandals. He was a mischievous child and as an adult was noted for his charm and his cunning. Zeus employed him as a spy to keep him informed of mortal affairs and took advantage of his untrustworthy nature to spread false rumours. Hermes also stole cattle from Apollo and became known as the patron of thieves. On a more positive

Snake charmer

The symbol of Hermes was the caduceus. This was a staff (originally an emblem of heralds) entwined with two snakes. According to legend, Hermes once separated two fighting snakes, which then entwined in peace.

Hermes is depicted wearing the winged helmet made for him by the blacksmith Hephaestos.

note, however, he was credited with the invention of the alphabet, astronomy, the lyre, mathematics and boxing. He was also believed to guide dead souls from Earth to the Underworld.

HESTIA

GODDESS OF THE HEARTH

Father: Cronos. **Mother**: Rhea. **Brothers**: Zeus, Poseidon, Hades. **Sisters**: Hera, Demeter. **Symbol**: Eternal flame. **Roman name**: Vesta.

The eldest daughter of Cronos and Rhea, Hestia was the goddess of the home and hearth. She was worshipped as the patron of the family and social order, and newborn babies were presented to her at the hearth. Pure, solitary and gentle in character, she never got involved in the quarrels among the other gods. She refused marriage to Poseidon and Apollo and never married. When Dionysus arrived at the court of Zeus she gave up her place to him, preferring to dedicate herself to tending the sacred fire on Mount Olympus. She was greatly respected by all the gods and Zeus allowed prayers to be said to her in any temple, regardless of the god to whom the temple was dedicated.

Sacred flames

Hestia's shrines at Delphi and Olympia were visited by thousands of pilgrims. Some took burning torches from her temples to provide divine protection for their homes in colonies overseas. Flames dedicated to Hestia were also kept burning in important public buildings.

POSEIDON

GOD OF THE SEA

Father: Cronos. **Mother**: Rhea. **Brothers**: Zeus, Hades. **Sisters**: Hera, Hestia, Demeter. **Children**: Amphitrite, Orion, Theseus, Polyphemus (a Cyclop), Pegasus (a winged horse) and others. **Symbols**: Trident, horses, dolphins. **Roman name**: Neptune.

While Zeus ruled the earth and the sky, his brother Poseidon was the god of the sea. He was said to live in an underwater palace and to roam the oceans in a gold chariot pulled by white horses. He could cause the seas to be calm or rough at will, using his three-pronged trident to whip up storms or tidal waves. Poseidon was also widely respected as the god of earthquakes and legend claimed he could sink whole islands or raise land from the sea bed. He was said to be easily angered and he quarrelled frequently with other gods.

SACRIFICES

Poseidon was depicted as a strong middle-aged man with long hair and a beard. Travellers prayed he would keep them safe on their journeys, and every two years the Isthmian Games sports festival was held in his honour at Corinth. Worshippers also sacrificed horses and bulls to him by throwing them into the sea.

The wrath of Poseidon

Among those who experienced the wrath of Poseidon was Odysseus on his voyage home from Troy. His ships were endangered by storms sent by the sea god.

The sea god Poseidon is depicted here seated and holding his characteristic trident.

LESSER GODS

Besides the Olympians, there were many lesser gods. Some were humans who had achieved immortality, while others represented various aspects of nature, such as winds and rainbows. Many were known only locally, but a few were familiar throughout Greece.

ASCLEPIUS

A son of Apollo, Asclepius was the god of medicine and healing. He was educated as a healer by the centaur Chiron. Athena gave him two flasks of Gorgon blood, one of which killed, while the other brought the dead back to life. Zeus, however, fearing this challenge to his power, struck down Asclepius with a thunderbolt. The sanctuary of Asclepius at Epidaurus in southern Greece attracted the sick from far and near in hope of a cure.

EOS

Eos was the goddess of the dawn. The daughter of the Titans Hyperion and Theia, she opened the gates of Heaven each morning to let the chariot of the Sun go through. She had four sons, who were called the Four Winds. They were Boreas (the north wind), Notus (the south wind), Eurus (the east wind) and Zephyrus (the west wind).

EROS

The son of Aphrodite and Ares, Eros carried a bow with which he fired magic arrows at both gods and mortals, causing them to fall in love. He was usually depicted as a winged child.

This famous statue of Eros, the Greek god of love, stands in Piccadilly Circus, London.

GLAUCUS

A minor sea god, Glaucus was born a mortal but he subsequently became a god after eating a magic herb. He could foretell the future and would rise from the waves to warn sailors of approaching danger.

HEBE

The goddess of youth, Hebe served up the ambrosia and nectar on which the gods lived at the court of Zeus until she was replaced by the beautiful youth Ganymede. When the Greek hero Herakles (see page 91) achieved immortality and entered Heaven, Hebe became his wife.

HELIOS

The brother of Eos, Helios represented the Sun. He was always depicted as a young man with rays of light shining from his head, riding the chariot of the Sun across the sky each day.

HYGIEIA

The daughter of Asclepius, Hygieia was worshipped as the goddess of good health, with her sisters Iaso and Panacea. Her temple at Epidaurus attracted large numbers of pilgrims.

IRIS

The goddess of the rainbow, Iris went on numerous errands for other gods, serving as their messenger. She also assisted mortals in need of help.

NEREUS

Otherwise known as one of the Old Men of the Sea, Nereus lived at the bottom of the ocean in company with his 50 beautiful daughters, who were called the Nereids. Identified as one of the Titans, he was noted for his truthfulness and virtue. He was also famed for his prophetic powers and for his ability to change his shape at will.

NIKE

The goddess Nike was the daughter of the goddess of the River Styx. Depicted as a beautiful young woman with wings, she helped Zeus alongside her siblings Striving, Strength and Power and rewarded victory in sport or war.

PAN

The son of Hermes, Pan had the upper half of a man and the lower half of a goat and was worshipped as the god of flocks and shepherds. A good dancer and musician, he invented the first pan pipes, which he made from the reeds into which Zeus transformed the nymph Syrinx when Pan pursued her. Pan's fondness for playing tricks on travellers inspired the word 'panic'. Worship of Pan was associated particularly with Arcadia, a district of rural mountain dwellers.

PRIAPUS

A son of Aphrodite, Priapus was a fertility god and a companion of Pan. Deformed in the womb by the jealous Hera, he was ugly and deformed and so lustful that he found it difficult to move. Statues of Priapus were sometimes placed in fields and gardens to ensure a good crop. He was a favourite comic character of ancient Greek poets and playwrights.

PROMETHEUS

Prometheus was a Titan who stole fire from Mount Olympus and took it to Earth hidden in a hollow fennel stalk. Zeus was outraged by the theft and sentenced Prometheus to be chained to a rock. Every day an eagle was sent to tear out his liver. After many years of this torture, Prometheus was rescued by Herakles, who shot the eagle and set Prometheus free. As well as punishing Prometheus for smuggling fire from Heaven, Zeus took revenge on mortal men by having Hephaestos fashion Pandora, the first mortal woman. According to legend, it was the guileful Prometheus who created the first man.

PROTEUS

One of the Old Men of the Sea, Proteus looked after Poseidon's seals. He could change his appearance into anything he wished. He was said to be a son of Poseidon and thus a brother of Nereus.

SELENE

The sister of Eos and Helios, Selene was a moon goddess who fell in love with the mortal Endymion. She persuaded Zeus to make Endymion immortal, but he imposed the condition that the youth would remain asleep forever. Selene agreed to this condition and spent each night gazing at her sleeping lover.

THEMIS

The second wife of Zeus, Themis was a Titan but she was allowed to organize feasts and other ceremonies for the gods on Olympus and was identified as the goddess of justice on Earth. She continued to serve Zeus even after he took Hera as his third wife.

TRITON

The son of Poseidon and Amphitrite, Triton had the upper half of a man and the lower half of a fish and shared his father's power over the waves.

COMPANIONS TO THE GODS

The gods of ancient Greek mythology had numerous companions of various kinds. Some were fantastical creatures who were part human and part animal, while others were attendants or nature spirits depicted as beautiful young men or women.

FATES

The Fates were three old women in whose hands the lives of mortals were measured and ended. The daughters of Zeus and the Titan Themis, they were Clotho (who spun the thread of mortal lives), Lachesis (who measured the length of each thread) and Atropos (who cut the thread).

FURIES

The Furies were three ugly hags who dealt out punishment to guilty souls in the Underworld. Called Tisiphone, Megaera and Allecto, they were said to have the heads of dogs, the bodies of old women, and bats' wings and snakes for hair.

GRACES

The Three Graces were the companions of Aphrodite, the goddess of love. Depicted as three beautiful maidens, they were identified as daughters of Zeus by the sea nymph Eurynome, and were called Aglaia, Euphrosyne and Thalia.

HORAI

The Horai were three nature spirits who were sisters of the Fates. Called Eunomia, Dike and Eirene, they cared for plants and looked after Hera's horses and chariot. They were goddesses of fairness and social harmony.

MAENADS

The Maenads were a group of 'wild women' who attended upon Dionysus. Clad in deerskins and wreaths of ivy or live snakes, they engaged in frenzied dancing and music and were rumoured to handle dangerous animals like leopards and wolves.

THE MUSES

The nine Muses were daughters of Zeus by Mnemosyne. Each was linked with a particular branch of the arts and might be called upon for artistic inspiration. Believed to live on Mount Helicon in Boeotia near the Hippocrene spring, the Muses were Calliope (poetry), Clio (history), Erato (lyric poetry), Euterpe (instrumental music), Melpomene (tragedy), Polyhymnia (mime), Terpsichore (dancing), Thalia (comedy) and Urania (astronomy).

NYMPHS

Nymphs were female spirits who were associated with nature and fertility. Most were semi-divine, having been born to a god or goddess (in most cases, Zeus) and a mortal partner. There were many categories of nymphs, including dryads (nymphs of the woods), leimoniads (nymphs of the meadows), naiads (nymphs of springs and streams), nereids and oceanids (nymphs of the sea).

Echo and Narcissus

One of the most famous stories related about the various companions of the gods was that of the nymph Echo, who was a follower of Zeus' wife Hera. Hera took offence at Echo's tactless chatter about Zeus' many love affairs and eventually took away Echo's voice, so that from then onwards she could do no more than repeat what others said. When Echo then fell in love with the handsome Narcissus (a vain youth who spent his time gazing lovingly at his own reflection), she was unable to tell him she loved him. Heartbroken, Echo pined away among the rocks until only her echoing voice was left.

A Roman story?

The story of Echo and Narcissus was essentially one of Roman origin, as the Roman poet Ovid was the first to combine the previously separate Greek legends about the two characters.

SATYRS

The satyrs were lusty spirits of mountains and woods, and like other followers of Dionysus spent their time drinking and revelling. Because they failed to guard Dionysus as instructed by the gods, Hera punished them by giving them a grotesque appearance, with the upper bodies of humans and the horns and lower bodies of goats.

A painting of a group of satyrs and maenads, clearly showing their grotesque physical appearance.

THE UNDERWORLD

The ancient Greeks believed that the souls of the dead were escorted by Hermes to the Underworld. On payment of a silver coin, they were rowed across the River Styx by Charon, the ghostly ferryman, to the land of the dead – mourners put a silver coin in a dead person's mouth for this purpose.

Good souls enjoyed everlasting bliss in a sunlit paradise called the Elysian Fields. However, bad souls endured eternal suffering in a cold and gloomy region known as Tartarus. Those who had been neither good nor bad were sent to Asphodel, which was a grey, boring place where souls drifted aimlessly in the shade. The dead souls were prevented from escaping by a three-headed dog called Cerberus who guarded the entrance to the Underworld, which was popularly supposed to be located at Avernus, a crater near Cumae.

HADES

The ruler of the Underworld was Hades (otherwise known as Pluto), the brother of Zeus and Poseidon. He was stern and grim and drove around his domain in a black chariot, which was drawn by black horses. His possessions included a Cap of Darkness, which made the wearer invisible.

Orpheus and Eurydice

Orpheus was the best musician in Greece. When his wife Eurydice died of snakebite, Orpheus followed her to the Underworld. Hades enjoyed his music and allowed Eurydice to go free on condition that Orpheus did not look back as they left. Orpheus, however, could not resist the temptation and glanced back, only for Eurydice to vanish forever.

Orpheus and Eurydice, imagined at the exact moment that they are separated from one another forever.

MYTHICAL BEASTS

Greek myth included a host of fantastical monsters who battled with heroes or terrorized mortals at the command of the gods.

CENTAURS

Centaurs had the head and upper body of a man mounted on the body of a horse. Some were wise and kindly, but

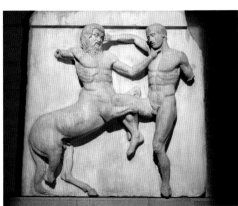

 This relief depicts the epic battle that took place between the centaurs and the Lapiths.

others fed on raw flesh and went mad at the sight of women or the smell of wine. Centaurs fought with the Lapiths (a peaceful mountain people) and with the hero Herakles, who killed them with poison-tipped arrows. They got their revenge by tricking Herakles' wife into giving him a poisoned shirt to wear, causing his death.

CHIMERA

The Chimera was a fire-breathing double-headed monster, with the heads of a lion and a goat, and a snake for a tail. Originally a pet belonging to the king of Caria, it escaped and ravaged the kingdom of Lycia until it was eventually killed by Bellerophon riding the winged horse Pegasus.

CYCLOPES

The Cyclopes were a race of giants who had a single eye in the middle of their forehead. The sons of Gaia and Uranos, they helped Zeus to defeat the Titans after he released them from the Underworld. They also served as blacksmiths to the gods.

ECHIDNA

The Echidna was a terrifying snake that became the wife of the giant Typhon. Together they became the parents of such monsters as the Chimera, the Hydra and the Sphinx.

GORGONS

The three sisters Stheno, Euryale and Medusa were the daughters of sea monsters and had tusks like boars, gold wings, bronze hands and live snakes for hair. Their mere glance turned men to stone. Medusa was killed by the hero Perseus, who avoided looking at her by using his shield as a mirror.

 This mosaic depicts Medusa, her gruesome head surrounded by twirling live snakes.

GRAEAE

These three sisters of the Gorgons, named Pemphedro, Deion and Enyo, were fierce, hideous hags who shared a single tooth and eye. Perseus eventually stole the eye to help him defeat Medusa.

GRIFFONS

Griffons were monstrous birds with the heads, wings and talons of eagles but the bodies of lions. Sacred to Zeus, they were said to live in lands north of Greece, guarding over hoards of gold belonging to the gods.

HARPIES

The Harpies had women's heads, the wings of birds and bodies of lions. They lived on islands in the Aegean Sea and fed on people's souls. When the Thracian king Phineus angered Zeus, he sent the Harpies to peck out the king's eyes and steal his food, and it was only when the Argonauts arrived in Thrace that the Harpies were eventually driven away.

HUNDRED-HEADED GIANTS

These three sons of Gaia and Uranos each had 100 arms and 50 heads. They sided with Zeus in his battle with the Titans.

SCYLLA AND CHARYBDIS

Scylla and Charybdis were two fearsome sea monsters who lived in a cave on opposite sides of the Strait of Messina. Scylla had six terrifying dog-heads that could tear sailors limb from limb. Charybdis created whirlpools that could suck down whole ships. Jason and Odysseus were among the heroes to confront them.

SIRENS

The Sirens were female sea monsters variously depicted with birdlike features or the heads of women. They lived on rocks onto which they lured passing sailors with their beautiful singing. To avoid such a fate, Odysseus filled the ears of his sailors with wax and had himself tied to the ship's mast so that he could not jump overboard.

SPHINX

The Sphinx had the head and chest of a woman, the body and tail of a lion and the wings of a bird. Anyone who failed to answer her riddle ('Which animal has at first four legs, then two legs, then three legs?') would be killed and eaten. When Oedipus solved the riddle, the Sphinx dashed herself to death on some rocks. The answer was 'A man', because a man crawls as a baby, walks as an adult and uses a stick when he is old.

HEROES

Alongside their myths concerning the gods, the Greeks had numerous stories about heroes, many of whom enjoyed semi-divine status as the son of a god and a mortal. Their remarkable and often bloodcurdling feats inspired many songs, plays and poems in both ancient times and in later centuries, and some were revered in much the same manner as the gods themselves.

The heroes of Greek mythology typically undertook arduous journeys in search of fabled treasures or to conquer evil in the form of some terrible monster. Many of these heroes had superhuman powers or they possessed magical gifts from the gods that helped them in their quests. A few, such as Herakles (opposite), eventually became gods themselves. The feats of such heroes were much repeated, and they helped to strengthen national pride.

BELLEROPHON

The hero Bellerophon waged war with various enemies, including giants and the Amazons. He also tamed the winged horse Pegasus, but when he flew too close to Mount Olympus, Zeus sent a gadfly to sting Pegasus. Bellerophon was thrown from Pegasus's back and lost his sight when he fell into a thorn bush.

HERAKLES

The greatest of the ancient Greek heroes was Herakles (Hercules to the Romans), the son of Zeus by the mortal Alcmene. Born a mortal, as a baby he strangled two huge snakes sent by the goddess Hera to kill him. He grew into an immensely strong, fierce and unruly man and, with the assistance of various gods, performed a series of miraculous feats called the Twelve Labours. These were set by King Eurystheus of Tiryns as punishment for killing his wife and children in a rage.

Herakles also loved wine and was noted for his many love affairs. When he died, he was carried by a cloud to Olympus where he was granted immortality and joined the ranks of the gods.

The Twelve Labours

1 KILLING THE NEMEAN LION
The first labour that was performed by Herakles was that of killing the fearsome Nemean Lion. Having achieved this arduous deed, Herakles was often depicted wearing the lion's skin.

2 KILLING THE LERNEAN HYDRA
For the second labour, Herakles killed the Lernean Hydra, a nine-headed dragon with poisonous breath; despite the fact that as each of its heads was cut off, two more appeared.

Herakles, adorned with the skin of the Nemean Lion, engaged in battle with the Lernean Hydra.

3 CAPTURING THE KERYNEIAN HIND

The third labour involved Herakles capturing a hind with golden horns that lived on Mount Keryneia. The pursuit took a whole year before Herakles finally trapped it in a net.

4 CAPTURING THE ERYMANTHIAN BOAR

The area around Mount Erymanthus was being ravaged by a giant boar. Herakles succeeded in capturing the boar and brought it back to Eurystheus, who was so terrified that he hid in an urn.

5 CLEANING THE AUGEAN STABLES

The vast stables of Augeas were full of horse manure that had gathered over the years. Herakles cleaned them by diverting two rivers through the stables.

6 KILLING THE STYMPHALIAN BIRDS

The Stymphalian Birds were terrifying man-eating birds that lived on a lake in the northeast Peloponnesus. Avoiding their poisonous feathers, Herakles scared them with a bronze rattle, which he had been given by Hephaestos, and then shot them out of their trees.

7 CAPTURING THE CRETAN BULL

The huge and ferocious fire-breathing Cretan Bull was the father of the Minotaur that was killed by Theseus. The bull ravaged Crete until it was taken captive to Tiryns by Herakles.

8 STEALING THE MARES OF DIOMEDES
Diomedes, king of Thrace, fed his mares on human flesh. Herakles, however, killed Diomedes, fed him to his own horses and then tamed them before taking them to Tiryns.

9 STEALING THE GIRDLE OF HIPPOLYTE
Hippolyte, queen of the fearsome Amazon race of female warriors, possessed a fine girdle that Eurystheus ordered Herakles to steal. Herakles accordingly killed Hippolyte and stole her girdle.

10 STEALING THE CATTLE OF GERYON
For his tenth labour, Herakles was ordered to steal the herd of red cattle belonging to the three-bodied monster Geryon. Herakles succeeded in killing Geryon, his giant herdsman and his hound and then took the cattle back to Tiryns.

11 STEALING THE APPLES OF THE HESPERIDES
The Hesperides were nymphs who owned a tree that bore golden apples. Herakles killed Ladon, the dragon that guarded the tree, and stole the fruit.

12 CAPTURING CERBERUS
For his final labour, Herakles captured the terrifying three-headed dog Cerberus, who guarded the entrance to the Underworld, and brought it to Tiryns, who fled in terror at the sight of the beast.

ICARUS

Icarus was the son of Daedalus, a famed craftsman and architect. After killing his nephew Talos in a rage, Daedalus fled Athens with Icarus and arrived in Crete. There, Daedalus designed the labyrinth of King Minos, but the pair were then imprisoned after Theseus killed the Minotaur. Daedalus and Icarus escaped using wings attached with wax, but Icarus flew too near the sun; the wax melted, and he was drowned in the Aegean Sea.

 Icarus plummets to his death after ignoring his father's advice and flying too near the sun.

JASON

Jason was a prince from Thessaly in northern Greece, who was sent on an impossible mission by his uncle Pelias after the latter assumed power. With his 50 companions, the Argonauts, Jason sailed north in the Argo in search of the magical Golden Fleece. After many perilous adventures, the Argonauts found the Fleece hanging from a tree at Colchis, on the Black Sea. With Athena's help, Jason killed the serpent that guarded the Fleece and returned home in triumph.

ORPHEUS

Orpheus was a poet and musician who sailed with Jason and the Argonauts. He used his music to calm rough seas and to maintain peace between the crew members. He also tried to rescue his wife Eurydice from the Underworld after she died from snakebite, but he lost her forever after he failed to obey the strict instruction not to look back at her as they returned to the world of mortals.

PERSEUS

Perseus boasted that he could kill the dreaded gorgon Medusa, whose very glance could turn a man to stone. Hermes gave Perseus some winged sandals, while Hades lent him a magic helmet and Athene gave him a polished shield. The winged sandals enabled Perseus to fly to the the lair of the Gorgons, where the helmet made Perseus invisible, so he could slip past Medusa's guardians, the Graeae. Using the shield as a mirror, Perseus avoided looking directly at Medusa and beheaded her.

 A statue of the hero Perseus holding the severed head of the dreaded Medusa.

THESEUS

The son of Aegeus, king of Athens, Theseus volunteered to be one of the tribute of seven young men and seven young women that Athens had to pay every nine years to King Minos of Crete as a sacrifice to be fed to the Minotaur, a monster that was half-man and half-bull. Theseus's courage won over Minos's daughter Ariadne, who told him the Minotaur lived in a labyrinth maze. She gave him a sword and a ball of thread, telling him to unwind it to enable him to find his way out. After a fierce struggle, he killed the Minotaur before following the thread he had unravelled back to the entrance.

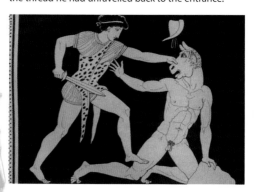

The Greek hero Theseus is depicted here slaying the legendary Minotaur, which was half-man, half-bull.

TEMPLES

The Greeks believed that the gods needed homes on Earth and built magnificent temples for them in the hope that they would be rewarded with divine favour and protection. Many temples contained a large statue of the relevant deity, placed where it could observe the outside world. The temples themselves were made in as magnificent a style as possible, using fine materials. Craftsmen were employed to create statues, columns, painted friezes and other decorations. The huge blocks for building temples were transported by ox-drawn carts and hauled into place using ropes and pulleys.

TEMPLE LAYOUT

The earliest temples comprised a single large room (the *cella*) and a pillared porch. Later temples were larger, with pillared porches at the front (the *pronaos*)

Altars

Ordinary people were not allowed into the temples to pray, as they are today in modern churches. Instead, a stone altar, where sacrifices could be made and offerings could be left, was usually sited outside a temple, in front of the main entrance.

and back (the *opisthodomus*). The largest temples, built in the Archaic Period, had a row of pillars (the *peristyle*) around the outside. They often included a treasury in which statues, jewellery and other offerings were stored.

THE PARTHENON

The temple of the Parthenon in Athens, dedicated to the goddess Athena, was constructed from marble by Iktinos and Kallicrates on the orders of Pericles in the fifth century BC. The blocks were fitted together with pegs, not mortar, while the columns were skilfully shaped to create the illusion that they were made out of a single piece of stone.

OTHER TEMPLES

Some of the finest temples built by the ancient Greeks remain as impressive ruins. The Temple of Hephaestos in Athens, known as the Theseum, is perhaps the most complete survival of all. Other notable temples included the Temple of Zeus at Olympia, which formerly housed a massive statue of the god, the temples of Hera, Apollo and Athena at Paestum, the Temple of Aphaea at Aegina and the Temple of Hera at Selinus. The Temple of Concord at Agrigentum is especially notable because it retains a complete peristyle. Other temples include the Temple of Poseidon on Cape Sunium, the Temple of Apollo Epikourios at Bassae and an unfinished temple at Segesta.

ORACLES

The ancient Greeks believed it was important to check that they had the favour of the gods before embarking on any major project. In order to do this, they might visit a soothsayer, who could foresee the future and would advise them on basic matters, such as deciding whom to marry or when to go on a journey. For advice on more complex problems, such as waging war on an enemy, wealthier Greeks or representatives of city-states went to a special temple called an oracle, where they paid priests or priestesses to communicate messages from the gods. At the oldest oracle, at Dodona in northwest Greece, messages from Zeus were interpreted through the rustling of the leaves in a grove of oak trees or through the cooing of doves in their branches.

THE ORACLE AT DELPHI

Most famous of all was the oracle at Delphi on Mount Parnassus, dedicated to Apollo. Messages there were conveyed through the Pythia, a priestess who lived in Apollo's temple and went into a trance in order to communicate with the gods. The Pythia breathed in smoke from burning leaves before answering any questions handed to her by her attendant priests. The priests then interpreted her often mysterious answers.

The centre of the world

The oracle at Delphi contained a stone, which was called the Omphalos. It was believed to mark the centre of the Earth. According to legend, when Zeus wanted to measure the world he set two eagles free at opposite ends of the Earth – Delphi was where they met.

OTHER ORACLES

The oracle at Dodona, which was dedicated to Zeus, Herakles and Dione, ranked second only to that at Delphi among the oracles of ancient Greece. Among other respected oracles further afield were those of Apollo on Crete and of Amun in Egypt, which was once consulted by Alexander the Great.

UNDERSTANDING THE GODS

Interpreting messages from the gods through the oracles was no easy business. When Croesus of Lydia visited the oracle at Delphi for advice about the wisdom of launching a military campaign against the Persians, he received the answer 'If you do, you will destroy a great empire'. Croesus took this to be a reference to the Persian empire and accordingly he ordered his troops to attack, only to find that it was his own empire that was destroyed.

CULTS AND RITUALS

Some ancient Greeks joined religious cults dedicated to the worship of a particular god or goddess. Members of these secretive groups had to train and keep to a virtuous way of life before they were fully initiated. The most famous cult was that of Demeter and Persephone at Eleusis (the Eleusinian Mysteries), which was notable for its annual initiation ceremony. This lasted several days, with sacrifices, acts of purification and a torch-lit procession from Athens to Eleusis.

RITUALS

Rituals in temples and public shrines were presided over by priests and priestesses, who also looked after statues of the gods and goddesses. Some were trained in the arts of prophecy and the interpretation of omens

Raised hands

Although much remains obscure about the religious practices of the ancient Greeks, we do know they prayed during ceremonies. In most cases, prayers to the gods were delivered with hands and arms raised. Prayers to Hades were made with palms towards the ground, while worshippers praying to Poseidon turned to face the sea.

(such as the flight of birds or thunder and lightning). Wealthy Greeks brought offerings of food, wine, incense, gold, silver and fine cloth and also paid for animals to be sacrificed. Poorer people brought pastries shaped like animals as they could not afford the real thing.

Most private houses had their own altar, which was usually situated in a central courtyard. Here small sacrifices would be made daily in the hope of winning divine protection for the whole household. On other occasions, a jar of wine might be poured over the altar to procure divine blessing.

This painting depicts a priestess of Bacchus, the Roman god of wine identified with Dionysus.

FESTIVALS

The ancient Greeks observed numerous festivals in honour of their gods and goddesses. They might feature theatrical performances and athletic competitions as well as religious ceremonies and grand processions. The whole population was given a holiday in order to attend the festival, which might last for several days, and food and drink was free. Some festivals were held annually; others might be held at four-yearly intervals.

ATHENIAN FESTIVALS

The people of Athens had over forty annual festivals, of which the most important was the Great Panathenaea, held in honour of the city's patron goddess Athena. It was held every four years and lasted for six days. It included a big procession in which crowds escorted a new dress for the massive wooden statue of Athena kept in the Acropolis. Another Athenian festival was the Anthesteria, held annually in the spring, when a statue of Dionysus was taken in procession to his temple and families laid food for the dead on household altars.

Processions

People participating in religious processions danced, sang hymns, played musical instruments, carried food and wine and led sacrificial bulls and other animals.

PART FOUR

Life in ancient Greece

Because the ancient Greeks left written records, which were rich in detail about the world in which they worked and played, we have a fairly comprehensive understanding of the lives they led. The study of ancient writings, combined with archaeological finds and the legacy of ideas passed down through the Romans and later civilizations, means we know a lot about ancient Greek society and the working and domestic lives of its people.

The School of Plato

The meaning of life and how it should best be lived was pondered by Greek philosophers, such as those of the School of Plato who are depicted here in an ancient mosaic.

GREEK SOCIETY

The population of ancient Greece through most of its history could be divided into two broad groups – free people and slaves. In the earliest times, most Greeks were poor, scraping a living as farmers or working as fishermen or craftsmen. Later, however, a new middle class emerged, consisting of merchants, craftsmen and moneychangers, and most menial work was done by slaves. The middle classes were the people who were to provide the foundation of Greek democracy.

FREE MEN

The most privileged free men were called citizens. Citizens were allowed to vote and take a full part in political life. In return, they were expected to fight in the army and to serve the city-state as government officials or as jury members. Citizenship was, however, only open to males aged eighteen or over: younger males, women, foreigners and slaves were all barred. In Sparta, moreover, only men of rich families were allowed to own property and take part in politics.

Citizenship in Athens

In Athens, citizenship was limited to the sons of Athenian parents. Men who settled in Athens but had been born elsewhere were called *metics* and obliged

Women

Women were not permitted to take any part in political or legal affairs. Their status in society depended entirely upon the standing of their husband or family.

to pay tax and perform military service for the state without being able to own land or property or to participate in political affairs.

SLAVES

Ancient Greek society was based on slavery. Slave traders made fortunes selling slaves as servants and labourers to wealthy Greeks. The ownership of slaves was a demonstration of wealth, and most well-off families had slaves to perform jobs around the house.

Work and duties

The duties of domestic slaves ranged from cooking and cleaning to looking after children and escorting them to school. Outside the home, slaves were also bought to work on farms, in mines, on building projects and as assistants to craftsmen, among many other roles. Some slaves were well-educated or highly-skilled and, as such, they might even be paid for their work. The cleverest ones were taken on as tutors to educate the sons of rich families at home.

 These citizens have gathered to buy slaves offered for sale by slave traders in the *agora*.

Harsh treatment

Slaves were bought and sold as the property of their owners and had no legal rights. Most were captured prisoners of war from newly-conquered lands. Others were foreigners purchased from slave traders. Some slaves were treated well and lived in wealthy households as members of the family. Others, however, were harshly treated, being kept in terrible conditions and watched over by soldiers as they worked.

Escape from slavery

Some slaves were paid for their work and could hope to save up enough to buy their freedom, although they would never be allowed to become fully-fledged citizens.

GOVERNMENT

During the Archaic Period, many city-states of ancient Greece, including those of Athens and Sparta, were ruled over by groups of rich local landowners, who were called *aristoi* (meaning 'best people'), hence the modern term 'aristocrat'.

OLIGARCHIES AND TYRANTS

Rule by a select group of powerful individuals (otherwise called an oligarchy) was sometimes replaced by that of a single man, or 'tyrant', who was chosen by the people. Some of these tyrants became notorious for their misuse of power, leading to revolts by the ordinary people and their replacement during the Classical Period by more democratic systems of government, the character of which varied from city-state to city-state. Sparta was the one exception to this, continuing to be ruled by a warrior oligarchy known as the Spartiates.

LAWS

Each city-state had its own laws, by which it governed its citizens. These laws tended to be stricter in military states such as Sparta and more liberal in city-states such as Lesbos and Athens, which, perhaps as a result,

Draconian laws

A set of new laws introduced in Athens in 621 BC by a lawyer called Draco proved so severe, with the death penalty being imposed for even minor crimes, that even today a strict ruling or law may be called 'draconian'.

appear to have had a relatively low crime rate. Murder and corruption were generally punishable by execution, while other serious crimes, such as the defilement of a temple, could be punished by banishment or by the loss of social privileges. Trials in Athens were heard before a jury of at least 200 citizens, who ruled on cases without help from professional lawyers and judges.

DEMOCRACY

Democracy – a system of government in which the ordinary people decide who will run the country – was an invention of the ancient Greeks, and it was first introduced in Athens in 508 BC. Meaning 'rule of the people', it ranks among the most significant legacies of ancient Greek civilization.

Athenian democracy

As practised in Athens, ancient Greek democracy allowed Athenian-born male citizens over the age of eighteen (though not women, foreigners or slaves) the right to speak and vote on important political issues at regular open-air meetings of the Assembly. This was a gathering of at least 6,000 citizens that took place every ten days on a hill near the Acropolis, called the Pnyx. Here, statesmen made speeches from a stone platform, which can still be seen.

Issues were placed before the Assembly by a Council of 500 officials, who were chosen each year by lot, and whose duty it was to draw up new laws and policies and to ensure the smooth running of the state. In wartime, decision-making powers were passed to ten elected generals called the *strategoi*.

Ostracism

Members of the Assembly could vote at a special annual meeting against any politician they did not like by writing his name on a fragment of pottery called an *ostrakon*. If more than 6,000 votes for 'ostracism' were gathered, then the politician in question was banished for ten years (as happened to the Athenian leader Themistocles).

Voting

Votes at the Assembly were made by simply raising a hand. Speakers were timed by a special water clock: when the last of the water dripped from a jar the speaker had to stop.

ATHENIAN POLITICIANS

The founders of Athenian democracy were celebrated among their contemporaries, and their ideas are still widely quoted and debated today.

ARISTIDES

Aristides was an Athenian statesman who was born in 530 BC. Nicknamed 'the Just', he was described by Plato (see page 187) as the only man in Athens who was worth admiring. As a political leader, he proved himself to be both fair and generous. He was ostracized after opposing Themistocles and spent five years in exile before being elected *strategos* against the Persians in 480–479 BC. He helped set up the alliance called the Delian League and remained a respected leader in Athens until his death around 469 BC.

CLEISTHENES

Cleisthenes was an Athenian aristocrat who belonged to a powerful family called the Alcmaeonids. He came to power in 508 BC after his family emerged as the victors after two years of civil war and subsequently made important constitutional reforms leading to the introduction of democracy in Athens. He also helped fend off the threat of invasion by the Persians.

DEMOSTHENES

Born in Athens in 384 BC, Demosthenes was one of the most celebrated orators of ancient Greece. He argued for stiff resistance against the invasion by Philip II of Macedon, his most famous speeches in this cause being written down as the *Olynthiacs* and *Philippics*.

 The famous orator Demosthenes addresses the Assembly in Athens.

He went into exile after being accused of stealing money from the state treasury but later returned to lead a revolt against the Macedonians in 322 BC. However, he was defeated and took poison.

PERICLES

The nephew of Cleisthenes (see page 116), Pericles was born around 495 BC and, as an adult, he established a reputation both as a military leader and a democratic politician. Widely believed to be incorruptible, he was an extremely popular leader, being re-elected *strategos* (wartime leader) of Athens for fourteen years in succession (443–429 BC).

A wise ruler

Pericles was a great orator and did much to improve life for the ordinary Athenian people, providing them with more paid jobs, free entertainment and beautiful new buildings, which included the Parthenon. He gave more people a say in political affairs and showed great cunning in international diplomacy, forming the Delian League of Greek states to resist Persia.

The Golden Age

His wise rule during the period that became known as the Golden Age of Athens attracted many celebrated artists, sculptors, actors, musicians, writers, scientists and thinkers to the city. He died of a fever in 429 BC.

SOLON

Solon was an Athenian legislator and magistrate who in 594 BC passed a series of new laws that reversed the harsh measures previously imposed by Draco and in so doing established a fairer system of government. He cancelled the debts of the poor in an effort to improve their lives and paved the way for democratic rule through reforms in legal, economic and political fields.

Other popular decisions included action to stop merchants exporting much-needed grain, which he used instead to feed the poor people of Athens. He also allowed relatively poor citizens to participate in the Assembly and enabled the middle classes to hold administrative posts.

THEMISTOCLES

Born around 524 BC, Themistocles was a controversial Athenian statesman who served as a general at the Battle of Marathon in 490 BC and later built up the fleet with which the Athenians defeated the Persians at the Battle of Salamis in 480 BC.

Having laid the foundations of the Athenian maritime empire, he fell foul of pro-Spartan rivals at home, and around 471 BC was ostracized and had to flee to Asia, where the Persians made him governor of three cities.

WORKING LIFE

As Greek civilization evolved, increasing numbers of people abandoned their former lives as simple farmers and fishermen to move to the new cities in order to pursue new careers as craftsmen, traders, soldiers, teachers and administrators.

FARMING

With its mountainous landscape, poor soil and harsh climate (too much rain in the winter and not enough in the summer), mainland Greece offered the ancient Greeks little good farmland. Most of the land that was suitable for farming lay near the coast or in sheltered inland valleys.

Food to supply the population came chiefly from small farms in such locations, each run by a single family, sometimes with the help of a few slaves. Women and children were also expected to help with the work. It was a hard life, and many farmers struggled to produce enough to feed more than themselves.

The farming year

Greek farmers ploughed in the spring, and again in the autumn, using oxen or donkeys to pull their ploughs. In many places, local communities shared oxen.

Divine aid

When times were hard, farmers prayed for help to Zeus and Demeter, the goddess of spring and summer.

The main crops included wheat, corn, barley and olives. Grapes were also grown to be harvested and trodden on to make wine. Most farmers also kept bees to make honey, as well as goats and sheep for their meat and cheese, which they flavoured with herbs.

The olive harvest

The olive harvest was an important time in the farming year. Farmers spread large cloths under their olive trees and then beat the trees with sticks to make the ripe fruit fall. Olive oil was widely used for cooking, for lighting and in beauty products. Such was the value of the olive harvest that anyone who uprooted an olive tree could be punished by law.

MARKETS

The produce grown by the farmers of ancient Greece was brought into towns and villages to be traded in the *agora*. People flocked to the open-air stalls erected in the agora to buy fruit, vegetables, cheese, grain, chickens, piglets, fish, eggs, wine and a host of other items, even slaves.

The agora was the bustling heart of Greek communities and the place where moneychangers, lawyers and others held business meetings. Deals between traders and their customers were regulated by various officials, who checked the quality of goods offered for sale and confirmed that the weights and measures used by traders were accurate.

A popular meeting place

The agora was a popular meeting place and it was also here that local craftsmen worked and set out their goods for sale. People seeking work gathered in the agora, in the hope of meeting someone interested in hiring their services. The crowds were often entertained by musicians, dancers and acrobats.

The shady colonnade and other buildings that surrounded the agora in Athens and other cities was where philosophers ran schools for their pupils. Other features of the agora included statues of gods and prominent political figures, as well as an altar and platforms on which traders could display their wares.

Relics

The best-preserved example of an ancient Greek agora is at Izmir in modern Turkey. A relic of the agora is the word 'agoraphobia', which refers to a fear of open spaces.

MONEY

Originally, the ancient Greeks came to market to exchange their produce for other goods, as money had not been invented yet. The first coins appeared in the seventh century BC, when they were introduced in Lydia, a kingdom of Asia Minor, and their use quickly spread throughout the Greek world. The earliest coins used by the Greeks were made of electrum, a mixture of gold and silver, although later coins were made of silver and, occasionally, gold.

Different coins

Each city minted its own coins, which were often adorned with distinctive symbols, often depictions of gods or heroes. Those of Athens featured an owl, which was sacred to Athena, the patron goddess of the city. During the Hellenistic Period, it became customary to depict rulers on coins, a practice that has continued into modern times.

Rates of exchange

As well as using coins to make trade and commerce easier, the ancient Greeks realized that it was possible to raise some additional revenue from the coinage by charging foreigners to exchange their own money for that of the state they were visiting. Coins also provided a convenient way of handling valuable gold and silver resources.

TRADE

The wealth of ancient Greece was founded on its flourishing trading links with the rest of the ancient world. Slaves, wine, pottery, oil, statues, silver, metalwork and the work of Greek craftsmen were carried on hundreds of ships sailing between city-states and further afield, to Egypt, Syria, Sicily and elsewhere. On return trips from the Black Sea and distant Mediterranean, Greek ships brought back cargoes of grain, upon which the city-states relied to feed their populations. Other imports included such goods as timber, linen, wool, dye, gems, spices, carpets and even elephants.

Wealth creation

Individual Greek merchants made fortunes from such commerce, and even the city-states themselves also benefited through the payment of custom duties which were charged on the movement of goods. Athens, in particular, emerged as a major trading centre, as did its rival Corinth.

Taxation

Much of the wealth of the city-states of ancient Greece came from taxes on trade and other economic activity. People with large fortunes were expected to donate money to fund drama festivals, equip naval vessels or pay for various public projects. Taxes were also paid on houses, slaves, livestock, wine and crops.

TRAVEL AND TRANSPORT

The roads in the towns and cities of ancient Greece were well paved and drained, with pavements for pedestrians, as proved by surviving examples in Crete, Corinth and elsewhere. Because of the mountainous terrain of the Greek mainland, however, there were relatively few roads between towns. Travel across land, on foot or perhaps leading a mule or packhorse, was therefore difficult. Wealthier people made their journeys in horse-drawn chariots.

 This painting on a Greek vase depicts a light horse-drawn chariot.

TRAVEL BY SEA

It was often much easier to travel long distances by sea instead of overland, particularly with most towns and villages being situated near the coast. Most ships stayed close to the coast because of the contemporary lack of knowledge about navigation, though some ventured far across the oceans, sailing as far north as Britain and east to India.

Sea voyages had their own dangers, with the risk of sudden intense storms blowing up, especially in the winter months. There was also the threat of capture by the pirates who lurked in the waters of the Aegean waiting to pounce on slow merchant ships carrying rich cargoes of oil, wheat, wine and other produce.

Before setting sail, many captains made sacrifices to the sea god Poseidon, asking him to protect their ships from harm. They might also consult an oracle for reassurance about the best time to leave harbour.

Greek shipping

Cargo ships of the ancient world were typically large and slow wooden vessels, usually with a single square linen sail and leather ropes. Merchant ships had a large hold suitable for transporting bulky cargoes.

FAMILY LIFE

Family life was considered more important in some city-states than in others. The Spartans placed little emphasis on the family, but other Greeks observed sophisticated social rituals and took care over the education of their children.

LOVE AND MARRIAGE

Most marriages in ancient Greece were arranged by parents, who sought matches with wealthy or powerful families. Girls were usually married off around the age of thirteen or fourteen, though their husbands were usually aged thirty or more. In Sparta, men were not allowed to marry until they were at least twenty.

Wedding ceremonies

Brides bathed in water from a sacred spring the day before their marriage. On the wedding day, the bride and groom put on white tunics, made sacrifices and feasted. At dusk, they went to the groom's house, accompanied by a noisy procession. The bride was greeted by the groom's mother and carried over the threshold by her new husband. The couple shared food at the family hearth and were showered with nuts, fruit and sweets. The bride was then led to the bedroom, with much laughter. The next day her family visited and gifts were exchanged.

Relaxed attitudes

The ancient Greeks had a relaxed attitude towards nudity and sex. Many statues were naked, and athletes always competed in competitions unclothed. Greek myths were full of racy details about the seduction of mortals by gods. Love between people of the same sex was also widely accepted.

THE ROLE OF WOMEN

Women in ancient Greece had fewer rights than men. They could not vote, hold positions of authority or own money or property. Daughters received little or no formal education and were married off as soon as possible. Anything they possessed on marriage passed to their husband. Those who failed to find a husband remained under the authority of their fathers or brothers.

Confined to the home

Married women were considered the property of their husbands and spent most of their time at home in the women's quarters (the *gynaecium*), where they busied themselves with spinning, weaving, managing the household and receiving female friends. They were rarely allowed out of the house unaccompanied by their husbands or a slave, with most of the shopping and other errands being performed by slaves. Their prime duty was the bearing and raising of children, especially boys (who could inherit the family wealth).

Divorce

Wives who proved unfaithful to their husbands or failed to produce a male heir could be divorced. All that was required was for the husband to make a formal statement of divorce before witnesses, upon which the wife was sent back to her family. If the wife wanted a divorce, she had to find a man to represent her as she had no legal status.

Poor women

Women in poorer households that could not afford servants cooked the meals and also undertook all the menial tasks, such as cleaning and making clothes. Curiously, poorer women were often freer than their rich counterparts. Poorer women were allowed to work and to go about the town where they lived without being accompanied constantly by slaves.

Spartan women

Although Spartan society was renowned for its harsh discipline and lack of luxuries, Spartan women enjoyed some advantages. In Sparta, they were treated more like the equals of men. They were less confined to their homes than women in other Greek city-states, and they were encouraged to take part in athletics in order to produce and rear healthy babies who would support the state as soldiers. Domestic duties were also considered less important.

Dancing girls

Some young women found employment as *hetaerae*.
The *hetaerae* were slave girls who entertained the men
at drinking parties with a combination of conversation,
music and acrobatic dancing.

CHILDREN

The ancient Greeks brought up large families, knowing
that many of their children would not survive into
adulthood – only half of all the babies born reached
the age of twenty. Parents had the right to abandon
sickly or female babies, leaving them in the open air to
die or to be adopted by others – they were sometimes
brought up as slaves.

Infants

Newborn babies were formally welcomed into the
household in a ceremony called the *amphidromia*. This
involved the child being carried around the hearth by
the female members of the household and then named.

Because of the risk of an early death, the fact that
any child reached the age of three was a cause for
celebration within the family. In Athens, the occasion
was marked at the festival of Anthesteria, which took
place every spring. Each child was presented on the

second day of the festival with a small jug symbolizing the end of its infancy.

Toys

Wealthier parents hired a nurse to look after their infants and provided their young children with their own furniture (including potties) and toys. These included such things as rattles, whipping tops, yoyos, hoops and sticks, board games and jointed dolls as well as toy soldiers and animals that were made of wood, clay, leather or rags.

If children died young they were often buried with their toys for company in the afterlife. Children were considered young adults at twelve and were expected to throw away their toys, dedicating them to Apollo (for boys) and to Artemis (for girls).

A Spartan childhood

Life for children in Sparta was hardest of all. The Spartans aimed to turn all their children into warriors and denied them an academic education or any luxuries, training them instead in handling weapons and getting physically fit. From the age of seven, Spartan boys lived in barracks, where they were allowed just one tunic, no covers for their beds and little food. They were also thrashed regularly to teach them endurance and respect for their elders.

EDUCATION

Many of the city-states of ancient Greece placed great importance upon the education of children, so that they would grow into responsible citizens. In Athens and elsewhere, schooling was provided for the male children of free citizens. Lessons had to be paid for, however, so the sons of poorer families received little or no formal schooling.

GREEK SCHOOLS

Boys began their education around the age of seven, and remained at school until aged fourteen or fifteen. Rich families employed private tutors or paid a slave (a *paidagogos*) to accompany boys to school and keep an eye on their behaviour.

Pupils were taught by three types of teacher. The *grammatistes* taught reading, writing and arithmetic.

Education of girls

The education of most girls was restricted to learning household skills, including spinning and weaving, though the girls of some wealthy families might also be taught to read and write.

The *kitharistes* taught the playing of an instrument (the lyre or the pipes) and poetry. The *paidotribes* taught dancing and athletics. Children studied in small groups, without desks, writing on wax tablets with a wooden pen called a *stylus*. Any mistakes could be rubbed out.

Education for most boys ended before they went for military training at the age of eighteen. A few, however, continued their studies under travelling teachers called Sophists, who taught the art of public speaking, or at the famous schools that were founded by Socrates, Plato, Aristotle and other philosophers to encourage informal academic debate.

The gymnasium

Pupils attended lessons and undertook physical training in a specially-built structure called a *gymnasium* (hence the word 'gymnastics'). The *gymnasium* comprised a large open space surrounded by shady colonnades, which provided suitable surroundings for lectures or discussions. Other activities that took place there included socializing and athletic training for the public games. Wrestling and ball games were undertaken in a building called the *palaestra*, which was an essential part of the complex. Coaching in physical exercise was the responsibility of officers called *paedotribae* or *gymnastae*. The famous philosophical schools of Greece were mostly based in *gymnasia*, including the Academy, the Lyceum and the Cynosarges in Athens.

HOUSES

Most ordinary homes in ancient Greece were built with mud bricks and whitewashed to reflect the heat of the sun. Roofs were covered with baked 'terracotta' clay tiles or thatch. Floors were constructed with stone or wood, sometimes decorated with elaborate mosaics made from small coloured pebbles. Small, shuttered, glassless windows let in fresh air when it was needed.

Poorer families tended to live together in just one large room. The houses of wealthier families, however, usually had more than one storey (reached by a wooden stairway), with several rooms on each. The rooms, which were large and cool, were usually arranged around a central open courtyard, which contained an altar, where prayers for the protection of the family would be said.

LIVING AREAS

The family's living rooms were usually situated on the ground floor, with the bedrooms and servants' quarters above. Men and women had separate living areas and spent relatively little time together. Husbands and wives never ate together except when they had guests. Kitchens had open fires, with smoke escaping through a hole in the roof. There might also be a bathroom, containing a terracotta bath. The grandest room in the

House-breakers

Because of their construction from sun-dried mud bricks, little remains of most ancient Greek homes today. Their construction also meant that they were relatively easy for criminals to break into. The Greek name for house burglars was 'wall-diggers'.

house was the *andron*, a dining-room where the men entertained their male friends.

FURNISHINGS

The rooms of most homes were quite simply furnished with essentials such as beds, tables, chairs and couches. Clothes and other belongings were stored in large chests instead of in cupboards. Tapestries made by the women of the house might be hung for decoration on the walls. Rooms were heated by charcoal braziers and lit by oil-burning lamps.

SPACIOUS HOMES

The houses of ancient Greeks living in urban centres were relatively large in comparison to those of other ancient cultures. The typical area of an average-sized Greek house in the fourth century BC was around 230 square metres (248 square yards).

A typical Greek house

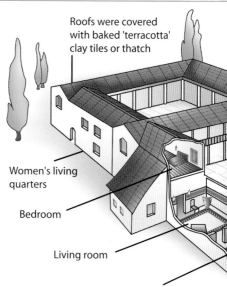

Roofs were covered with baked 'terracotta' clay tiles or thatch

Women's living quarters

Bedroom

Living room

Andron - dining room where men entertained their friends

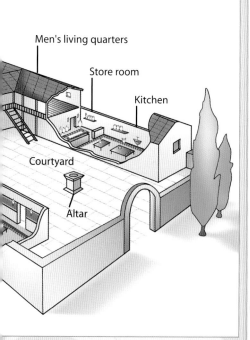

Men's living quarters

Store room

Kitchen

Courtyard

Altar

FOOD AND DRINK

The ancient Greeks enjoyed a simple but healthy diet, with plenty of fruit and vegetables, seafood, olive oil and honey instead of sugar. Other ingredients included bread, cheese, eggs, cakes, salads, spices and meat, which was usually mutton. Every part of the animal was used, including the lungs (which were fried), the intestines (which were stewed) and the brains (which were boiled).

The poor could rarely afford meat and lived mainly on bread and porridge, although they had the option of trapping and eating small birds such as thrushes and swallows. On other occasions, they might receive a share of the meat that was distributed among those present when an animal was sacrificed.

The Greeks also hunted for the table with dogs, their prey including hare and wild boar, and they foraged for nuts, berries, wild mushrooms and herbs.

DAILY MEALS

Most Greeks began the day with a small breakfast of bread soaked in wine, perhaps washed down with some milk or wine and water. Lunch was again fairly modest, as the main meal of the day was eaten in the cool of the evening. For poorer people, this was based on barley porridge or bread and vegetables.

The rich ate better food and the master of the house might hold a dinner party to which he invited male friends. Served in the family dining room, the meal might consist of several courses, accompanied by wine diluted with water. The food itself was cooked over an open fire or in clay ovens by the women of the house, or else by female servants. Between courses, diners snacked on sweetmeats (made from dates, nuts, figs, sesame seeds and honey). The meal was often followed by music and dancing or more drinking. Women were not invited to dinner parties held by men, but might sometimes invite their female friends to their houses to eat with them.

Wine

Wine was made after the grape harvest, the grapes being trodden on barefoot in large wooden tubs before being strained for drinking.

Table manners

Diners ate meals from low tables as they lay on couches. They used pottery plates and either a metal spoon and knife or their fingers. Men usually dined separately from the women, who ate with the children.

CLOTHING AND HAIRSTYLES

Due to the hot climate, both sexes wore loose, flowing robes to stay cool. Men wore a wool or linen tunic and over this a rectangular piece of material called a *chiton*, fastened with pins over one or both shoulders and belted round the waist. In cold weather they draped themselves in a cloak called a *himation*. Women wore a sleeveless garment (a *peplos*) or a long-sleeved chiton and himation. Children of both sexes wore short chitons.

FASHION

The ancient Greeks were very fashion-conscious and prided themselves on appearing more elegant than their neighbours. The tunics of the wealthy were often coloured with natural dyes and embroidered. Most clothes were made of wool or linen, although the rich might also wear silk or cotton from India and the East, sometimes so finely spun they were almost transparent.

Footwear

Outdoors, most people wore leather sandals, sometimes tied with long leather laces, or boots. The best shoes had thick cork soles, which made walking on stony ground more comfortable. Poor people mostly went barefoot.

Slaves usually wore plain woollen tunics and might work wearing just a loincloth and a shady hat.

HAIR AND BEARDS

After the Archaic Period, it became fashionable for men to wear their hair short, often with a beard. Greek women usually wore their hair long, often tied up in knots and coils and held in place with ribbons, pins, nets or headbands.

 The ancient Greeks loved to dress fashionably in loose flowing robes and tunics.

MAKE-UP AND JEWELLERY

Greek men and women devoted much effort towards making themselves attractive. They bathed regularly, washing themselves with water from a large pottery bowl and then scraping their skin clean with olive oil, using a specially-shaped tool called a *strigil*. They also washed their hair and rubbed perfumed oils into their skin to keep it moist in Greece's hot, dry climate. Women also used hair dyes, wigs and even wore shoes with thick soles to make themselves appear taller. Men cut and curled their hair and grew or shaved their beards in response to the fashions of the day.

MAKE-UP

Greek women preferred to have pale skin and often used white make-up to achieve this. They also used rouge to add pink highlights to their cheeks and black preparations to darken their eyebrows. As they applied their make-up, wealthier women could check their appearance in mirrors of polished bronze.

White lead

The make-up that women used to whiten their skin was often made of powdered lead, a highly toxic substance.

Jewels for the dead

The Greeks often buried valuable pieces of jewellery with the dead, with the result that many such items have since been recovered by archaeologists and can now be seen in museums around the world.

JEWELLERY

Both Greek men and women wore jewellery, often to indicate their wealth. The oldest items of jewellery were made of beads, which were sometimes fashioned into the shape of animals and shells. The first use of gold and gems appears to have been around 1400 BC. Both sexes used pins and brooches to fasten their clothes, while wealthier people might add various pieces of jewellery made of gold, silver and other valuable materials, sometimes decorated with precious or semi-precious stones, or enamel. These included elaborate necklaces, bracelets, earrings, pendants, rings, chains and headbands. Poorer people wore jewellery made from bronze or pottery.

Most pieces of jewellery fashioned by Greek craftsmen were carefully hammered from sheets of precious metal, sometimes based on designs borrowed from foreign cultures. More rare were pieces that were cast from special moulds.

LEISURE ACTIVITIES

The ancient Greeks enjoyed a wide range of leisure activities, which ranged from major public festivals and theatrical performances to private parties and entertainments in their own homes.

FESTIVITIES

The populations of whole cities would turn out for public pageants and festivals. These were usually free, being paid for by wealthy individuals or politicians who were seeking public support.

Women did not usually attend such festivities, however, and were similarly barred from attending the drinking parties (*symposia*) where the men gathered to eat, drink and discuss the affairs of the day. These social gatherings usually took place in private homes in the *andron*, which was set aside for the purpose.

Drinking games

Guests at drinking parties in ancient Greece were known to enjoy a game called *kottabos*. This involved drinkers wielding their cup by the handle and tossing the dregs of their drink at a target to see who could score the best hit.

Cockfights

The ancient Greeks also enjoyed cruel cockpit fights, setting animals such as cocks, quails or even cats and dogs against each other, so that they would fight to the death.

FUN AND GAMES

The ancient Greeks loved theatrical entertainments and they often hired musicians and dancers to perform at their parties, as well as singing and playing musical instruments themselves. Other forms of entertainment ranged from hunting and fishing to playing board games, some of which resembled modern chess or draughts. Women enjoyed playing knucklebones, a game that resembled modern jacks in which playing pieces were the ankle joints of small cloven-footed animals.

Physical exercise

Physical exercise was considered a vital part of every person's moral and ethical well-being, and also essential to good health. Popular pastimes of the more physical variety included wrestling, boxing and swimming. Many of these activities took place in the *gymnasium*, which was a feature of most major settlements. Men engaged in such activities naked, so (with the exception of Sparta where women were also encouraged to take part) only males were permitted to participate.

SPORT

The ancient Greeks believed that daily exercise was important and that a healthy body contributed to a healthy mind. Sport was popular throughout ancient Greece, especially in Sparta, where athletics was considered an important feature of military training. Children who were suspected of not trying hard enough were likely to be beaten by their teachers.

SPORTING FESTIVALS

Sporting festivals that were mounted in honour of the gods were held locally on a regular basis, but there were also bigger gatherings that attracted athletes and spectators from further afield. The sporting events included running races (over a range of distances) and competing in long jump, boxing, chariot-racing, wrestling and discus throwing.

Sporting highlights

The most important of the many sporting festivals and events that were held in ancient Greece was the so-called Panhellenic Games. These included the following four major festivals: the Isthmian, Nemean, Olympic (see opposite) and Pythian Games.

Male competitors at sports festivals competed naked, and women were not generally allowed to attend. Female athletes could, however, participate in their own sporting festival in honour of the goddess Hera, although only in running competitions.

THE OLYMPIC GAMES

The oldest and most famous sports festival of all was the Olympic Games, which was first held around 776 BC. Staged over five days on a four-yearly basis at a complex of temples and stadia at Olympia, the Games were part of a religious festival in honour of Zeus.

They began with a day of religious ceremonies and continued on the second day with horse-racing, chariot-racing and the pentathlon (comprising long jump, running, discus, wrestling and javelin). The third day of the Games featured more religious ceremonies and events for youths. The fourth day witnessed wrestling, boxing and track events, notably racing in armour. The festival ended on the fifth day with religious ceremonies and banquets.

The rewards of victory

Initially, at least, athletes did not formally represent their city-states at the Games, although victory was a source of great pride to their fellow-citizens. Winners were rewarded with a simple crown of olive leaves,

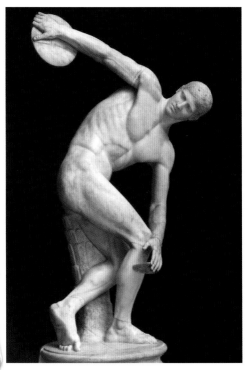

A Roman copy of The Discus Thrower, which was made by the Greek sculptor Myron.

The marathon

The long-distance marathon race is a feature of the modern games only. It was inspired by the Greek soldier Pheidippides, who ran 42 kilometres (26 miles) to bring news of the victory at the Battle of Marathon to Athens in 490 BC – and then died of exhaustion.

although the most successful sportsmen could earn a fortune from their sponsors and might pursue careers as professional athletes.

The Games were watched by huge crowds who came from all over Greece. Wars were suspended for a month before and after, so that athletes and spectators could travel to and from the festival. The Games were banned by the Romans in 393 AD, but later revived in 1896.

Olympia

The complex at Olympia, traditionally identified as the home of the gods, included a number of magnificent temples and other buildings. These were destroyed in 426 AD but have since been excavated by archaeologists. Among them were the Temple of Zeus, which was the largest Doric temple in Greece, the Temple of Hera, the aisled Echo Hall and the Stadium itself, as well as various facilities for the athletes and spectators. The Stadium could seat 7,000 people.

DEATH IN ANCIENT GREECE

Despite their relatively healthy lifestyles and advances in medical knowledge, few ancient Greeks survived beyond their forties. Many young men were killed in battle, while women often died in childbirth. Disease claimed many more lives, with thousands falling victim to the plagues that swept the ancient world.

FUNERAL RITUALS

The ancient Greeks believed that a person had to be properly laid to rest before their soul could be freed from the world of the living. The corpse of a dead person was washed carefully, sprinkled with perfumed oils and herbs and dressed in white, with the face left uncovered. A coin was then placed inside the dead person's mouth to pay the ferryman Charon who, it was believed, would row them across the river Styx, which separated the land of the living from the Underworld.

Grave paintings

The dead were often buried with their armour or with various treasured possessions. Food and other offerings might also be left in the grave in order to sustain the deceased after death.

Funeral processions

After a short time of mourning, the body was carried on a cart to a cemetery outside town. The funeral procession, which took place before dawn, was accompanied by grieving relatives, who dressed in black and cut their hair short. Women mourners sang funeral laments as the body was burned, buried or placed in a stone coffin. The graves of the rich were often marked by elaborately carved stone slabs which were called *stelae*.

Funeral orations

A speech praising the virtues of the dead person and urging those left behind to imitate their good deeds in their own lives was an essential feature of most funerals in ancient Greece. This was particularly true of big public funerals commemorating those who had died in battle, when the oration (*epitaphios logos*) was delivered by a prominent citizen – a great honour for the person selected.

Among the most celebrated funeral orations of all were those delivered by Pericles in honour of the Athenians who had died fighting in the Peloponnesian War, and by Demosthenes to commemorate the Athenians who died opposing the Macedonians in 338 BC.

Overleaf: Charon accepting payment to ferry a dead soul across the Styx.

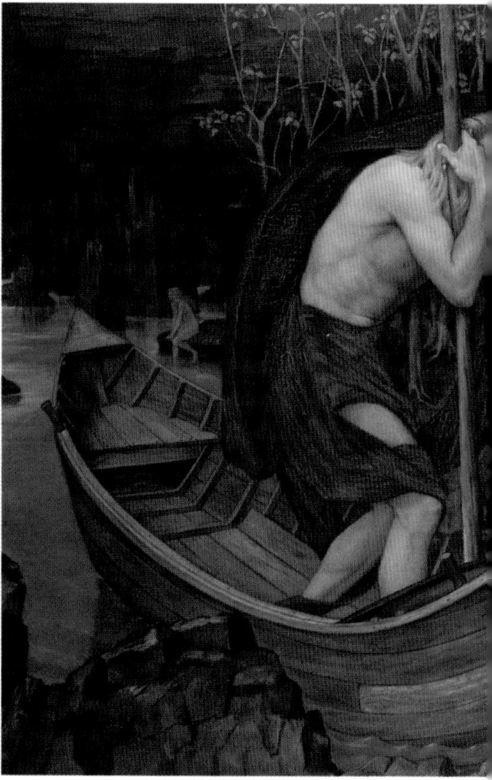

PART FIVE

Greek culture

The culture of the modern Western world has its roots in that of ancient Greece. The influence of the ancient Greeks is still profound in most cultural spheres, from architecture and literature to politics, philosophy and science.

The legacy of Greek culture

The cultural achievements of the ancient Greeks were to become a source of inspiration in later centuries.

CULTURE AND LEARNING

The greatest glory of ancient Greek civilization was its culture, which encompassed a wide range of activities, from arts and crafts to architecture, philosophy, politics, medicine, science, music and theatre.

ARTISTS AND CRAFTSMEN

Greek artists and craftsmen developed sophisticated ideas about what constituted good art and introduced the concept of mathematical proportion through the many masterpieces they produced. The most talented artists and craftsmen travelled from place to place to work on carvings, sculptures and new buildings. At a lower, more local level, thousands of metalworkers, basket makers, jewellers, leatherworkers, shoe makers, glassblowers, weavers and a host of other craftsmen also contributed to the building of the greatest civilization the world had seen.

WRITERS AND PHILOSOPHERS

Many of the most revered names associated with the culture of ancient Greece were writers or philosophers. Writers such as Homer preserved the great stories of their civilization, while others provided material for the Greek stage and in so doing inaugurated world theatre.

The Museum

The ancient Greeks established many important cultural centres. The most famous of these included the Temple to the Muses at Alexandria in Egypt, called the Museum. Together with its fine library, the Museum attracted scholars from throughout the Greek world.

The great Greek philosophers founded schools for the discussion of their ideas and greatly advanced thinking about government, science, medicine and religion among many other fields. Many modern assumptions about human society and behaviour ultimately date back to the ancient Greeks, and numerous scientific discoveries they made remain valid even today.

Intellectual curiosity

The intellectual curiosity that powered ancient Greek civilization extended to every aspect of the world the Greeks saw around them. They developed theories on the origins of the earth, drew maps of countries they visited, wrote histories of the known past, interested themselves in the properties of materials and natural processes, and drew conclusions about the nature of the universe from their observations of the heavens. They studied the character of man and deliberated on the intricacies of human society, founding new schools of thought to explain the workings of the human mind.

ARCHITECTURE

The homes of most Greeks were made of mud and brick and were fairly modest in design. The best architecture was reserved for temples and other important public buildings paid for by the city-states. The construction of a great building required the work of many skilled craftsmen, including architects, stonemasons and carpenters, as well as slaves.

PUBLIC BUILDINGS

There were many different types of public building. The *tholos* was a round building with a conical roof and columns round the outer wall. The *stoa* was a long building with a row of columns in front, often housing shops or offices. Treasuries resembled small temples and housed offerings to the gods. Other constructions included *propylae* (gateways to religious complexes), theatres and stadia, as well as massive altars and monuments to great heroes or military victories.

CONSTRUCTION

Temples and other major buildings were made of limestone or marble and (in the western colonies) sandstone, while roofs and ceilings were often made of wood, although roof tiles of terracotta (mixed clay

Mathematical symmetry

Greek architects believed that buildings should have a symmetrical, balanced appearance based on mathematical ratios. The height of the frontage of the Parthenon, for instance, was exactly four-ninths of its width.

and sand) or stone were also used. The blocks of stone were carved on the site with hammers, mallets and other tools, and laid in place using ropes and pulleys, then polished. Important buildings were finished with columns, statues and stone carvings called friezes. The walls inside might be decorated with painted murals, as were the houses of some wealthy families.

Doric temple frontage

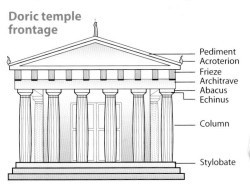

Pediment
Acroterion
Frieze
Architrave
Abacus
Echinus

Column

Stylobate

COLUMNS

The ancient Greeks did not know how to build arches or curved roofs and relied on vertical columns to support lintels upon which they could build sloping roofs. Most important buildings had dozens of columns in rows, each positioned with mathematical exactness in relation to their size to achieve a harmonious balance. The width of each column of the Parthenon, for instance, is precisely four-ninths of the distance between them.

CAPITALS

Each column was topped with a separate section called a capital, upon which the lintel rested. There were three main styles: the earliest was the Doric, with a plain, undecorated capital. In the Ionic style, the capital was more elegant, with a scroll-topped decoration called a *volute*. Later came the Corinthian style, in which capitals were decorated with acanthus leaves and other designs (although this is more associated with the Romans).

Caryatids

Most remarkable of all the columns put up by the ancient Greeks was the caryatid column, which was carved in the shape of an elegant woman. Famous examples form part of the Erechtheum on the Acropolis in Athens.

The caryatids of the Acropolis. They supposedly
represented the women of Caryae.

SCULPTURE

The ancient Greeks were master sculptors and they fashioned statues as monuments or to adorn temples, graves and homes. Some masterpieces may still be seen, while others are known through Roman copies.

Greek statues were made of wood or terracotta (mixed clay and sand) as well as stone, bronze, gold and silver. Early masterpieces included marble *kouros* (youth) and *kore* (maiden) figures in which subjects were depicted in stiff, formal postures, with one foot advanced and arms straight. Later sculptors such as Myron and Pheidias, however, depicted gods and mortals in more relaxed, sophisticated poses: running, sitting, lying down, riding horses or even fighting with one another.

MAKING A STATUE

Stone statues were carved with a mallet and chisel and then painted (though most paint has since worn off).

A living statue

According to Greek legend, a sculptor named Pygmalion made a statue so lifelike he fell in love with it. Aphrodite took pity on him and brought the statue to life.

The Venus de Milo, a masterpiece that is thought to depict Aphrodite.

Some figures were finished with eyes of glass, coloured stone or ivory, while bronze might be used for weapons, crowns and other details. Bronze statues were often made by the 'lost wax' method, in which a wax figure was encased in clay and heated so that the wax melted, to be replaced with molten bronze. Greek sculptors also carved stone slabs (called reliefs) to be mounted on temple walls. These depicted battles, processions and other scenes.

Destruction of statues

Many sculptures of pagan gods were destroyed during the early Christian era. During medieval times, numerous marble figures were lost when burned to produce lime, while bronzes were melted down for their raw materials.

POTTERY

Thousands of examples of ancient Greek pottery are preserved in museums. Using local clay, ancient Greek potters shaped pots on a revolving wheel and then fired (baked) the result in an oven heated by burning wood or charcoal. Most potters worked in quite small workshops close to their homes and were assisted by other family members or slaves. Many pots were signed by the potter and the artist who painted the pot after it had been fired.

USES AND SHAPES

Most pots were intended for everyday purposes. As well as jars and vases, potters also made cups, bowls, lamps and roof tiles. The shape of a jar or vase was dictated by its use. An *amphora*, for example, was a wide-bodied jar with two handles, and was used for storing wine or oil, while a *krater* was a large vessel in which wine was mixed with water.

Potters' quarter

Athenian potters (the most skilled of all) lived in their own special quarter of the city, called the Keramikos.

Other shapes included the *oinochoe* (a jug for pouring wine), the *hydria* (a jar in which water was fetched from a fountain) and the *skyphos*, *kylix* and *kantharos* (drinking cups). Some potters also made novelty pots in fantastic shapes.

DECORATED POTS

The earliest Greek pots were plain or bore simple geometric patterns. Later pots, however, were decorated with paintings of animals, plants, gods, heroes or everyday scenes. These were usually painted in black against a red background or, alternatively, in red against a black background.

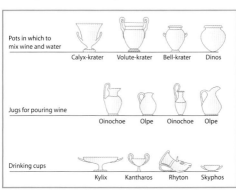

| Pots in which to mix wine and water | | | |
| Calyx-krater | Volute-krater | Bell-krater | Dinos |

| Jugs for pouring wine | | | |
| Oinochoe | Olpe | Oinochoe | Olpe |

| Drinking cups | | | |
| Kylix | Kantharos | Rhyton | Skyphos |

METALWORK

The ancient Greeks were highly skilled metalworkers. The Mycenaeans made weapons and tools from bronze and were famous for the fabulous death masks of beaten gold made for royal burials. Later generations used bronze, but they also made objects from iron.

BRONZE

The Greeks mixed tin and copper to make bronze, which they used for products ranging from mirrors and kitchen implements to armour and statues. The tin was imported from southern Spain, France and Cornwall, while the copper came from Greek mines or was brought in from Cyprus and elsewhere in the eastern Mediterranean. The ore was melted down in furnaces of baked clay bricks, heated by charcoal. The work was dirty and dangerous and each blacksmith was usually assisted by a gang of slaves.

IRON

Iron was introduced around 1050 BC. Weapons and tools made of iron were harder than those made of bronze. To make iron, blacksmiths had to raise the temperature in their furnaces. The iron was removed while still hot and hammered to remove impurities.

Athenian metalworkers

Athenian metalworkers worked in their own quarter near the temple of Hephaestos, the blacksmith god, mostly in workshops in their own homes.

GOLD AND SILVER

The precious metals gold and silver were usually reserved by the ancient Greeks for their most valuable objects, such as coins and jewellery. Silver was mined by slaves at Laurion near Athens from the eighth century BC onwards, or perhaps even earlier. The slaves who worked in the mine had to endure the harshest conditions, working shifts of up to ten hours.

Mycenaean death mask, which was once thought to be the gold mask of King Agamemnon.

LANGUAGE

The inhabitants of the various city-states that made up ancient Greece all spoke the same language, although there were several dialects (regional variations). The fact that they shared a common language was very important to the development of their civilization, and the influence of the language they spoke remains profound upon many languages of the modern world. Even the word 'alphabet' is of Greek origin. Opposite is a table showing the Greek characters and names and their English equivalents.

THE ALPHABET

The Greeks developed their written alphabet around 800 BC, basing it upon the earlier Phoenician alphabet, to which they added letters representing vowels. The Greek alphabet was easier to learn than earlier ones and many modern alphabets are based upon it.

Obsolete letters

Some letters of the ancient Greek alphabet fell from use at a relatively early date or were limited to certain dialects only. Some, like digamma, disappeared because they represented sounds that were no longer used.

Greek letters

Greek capital	Lower case	Greek name	English equivalent
Α	α	alpha	a
Β	β	beta	b
Γ	γ	gamma	g
Δ	δ	delta	d
Ε	ε	epsilon	e
Ζ	ζ	zeta	z
ΕΨ	εψ	eta	ey
ΤΗ	τη	theta	th
Ι	ι	iota	i
Κ	κ	kappa	k
Λ	λ	lambda	l
Μ	μ	mu	m
Ν	ν	nu	n
ΚΣ	κσ	xi	ks
Ο	ο	omicron	o
Π	π	pi	p
Ρ	ρ	rho	r
Σ	σ	sigma	s
Τ	τ	tau	t
Υ	υ	upsilon	u
Φ/ΠΗ	φ/πη	phi	f/ph
ΧΗ	χη	chi	ch
ΠΣ	πσ	psi	ps
ΟΗ	οη	omega	oh

LITERATURE

The Greeks had a strong storytelling tradition, though
for many centuries they lacked an alphabet to write
those stories. Professional poets called bards recited
tales about gods and heroes, some in the form of long
('epic') poems. These bards travelled from place to
place in search of audiences, and often accompanied
themselves on an instrument called a lyre (hence the
word 'lyrics' for lines set to music). The best bards were
widely respected and well paid for their performances.

GREEK POETS

Ancient Greece produced numerous celebrated poets,
although only fragments of their work survive. The
most famous of them included Hesiod, whose *Works
and Days* included practical advice about farming, and
the Athenian poet Pindar, who wrote seventeen books
on various themes, including victory in sport.

SAPPHO

Another remarkable poet was Sappho, a woman who
in the seventh century BC became famous for her
beautiful love poetry. She was born on the island of
Lesbos, probably into a rich family, and lived there most
of her life. She addressed love poems to the goddess

Aphrodite and other women (hence the modern word 'lesbian'), but is said to have killed herself when rejected in love by the boatman Phaon.

HOMER

The two epic poems that are known as *The Iliad* and *The Odyssey* were composed in the ninth century BC by a celebrated Greek bard called Homer, although they were only written down centuries after his death.

Little is known about the poet and his life, although legend claims he was born on the island of Chios and also that he was blind. Some have suggested he may have been more than one person, or a woman.

The first great epics

Homer's *Iliad* and *Odyssey* are the earliest surviving examples of ancient Greek literature, and are still revered as the first epic poems in European literature. They tell the story of the Trojan War and its aftermath as the warriors of Greece return home.

Written classics

It would take 24 hours to recite both of Homer's great works. One theory has it that the Greeks invented their alphabet solely for the purpose of recording the poems.

 This painting portrays a rapt audience listening to a reading of Homer's epic works.

Homer's admirers

The poems, which are expressed in beautiful and powerful language, were greatly admired by later generations, and Alexander the Great himself carried the works of Homer with him when he went to war.

Fragments of history

Many modern critics believe that the poems were probably not the work of Homer alone but a compilation of the work of many other poets over hundreds of years. It seems probable that the poems were based on fragments of history combined with legend and the poet's own imagination.

The Iliad

Homer's epic poem describes the last weeks of the ten-year Trojan War, which broke out between the Greeks and rival Troy after Helen, the beautiful wife of Menelaus, king of Sparta, was carried off by Paris, son of Priam, the king of Troy.

The poem particularly concerns the feud between the Greek warrior Achilles and the Trojan hero Hector. Achilles refuses to fight after falling out with the Greek leader Agamemnon, but in his absence his best friend Patroclus wears his armour and is killed by Hector, who mistakes Patroclus for Achilles. The remorseful Achilles then kills Hector and insults him by dragging his body behind his chariot three times round the walls of Troy. The poem ends with Hector's body being returned to Priam for honourable burial.

The Odyssey

The Odyssey begins with the final defeat of the Trojans after the Greek hero Odysseus suggests a way of getting Greek warriors secretly into the besieged city.

The Greeks build a huge wooden horse in which they hide some of their best warriors, before pretending to dismantle their camp and sail away. The Trojans think the horse is a gift to the gods and drag it inside the city, only for the hidden warriors to burst out during the night and let in the returning Greek army.

The city is sacked and Odysseus and his friends begin their journey home. They have offended the gods, however, and it takes the heroes ten years to arrive, after many adventures, which include encounters with the one-eyed Cyclops, the Sirens (whose lovely singing lures sailors to their deaths) and other deadly threats.

Penelope, the wife of Odysseus, has, meanwhile, been avoiding marriage to a host of men who believed that her husband must be dead. Odysseus slaughters all his rivals, and the lovers are reunited.

The Wooden Horse being pulled into Troy, a crucial episode near the beginning of Homer's *Odyssey*.

THEATRE

One of the most important contributions of ancient Greek civilization to world culture was the development of theatre. Most cities in ancient Greece had their own amphitheatre, in which a range of different types of drama was presented.

TYPES OF DRAMA

Types of drama included tragedies (sad, usually violent, stories that ended unhappily, often based on mythology) and comedies (happier, often very rude, tales that might poke fun at well-known public figures). Another form of drama was the satyr play – a comedy on a tragic theme in which the performers dressed as satyrs (half-man and half-animal).

DIONYSIA

The first plays emerged around the sixth century BC from the songs and dances that were performed in honour of the god of wine, Dionysus, and they retained their religious significance. There was even an altar on the stage where sacrifices could be made. Prizes were

Overleaf: The well-preserved theatre at Epidaurus, with 55 rows of seats capable of seating 14,000 spectators.

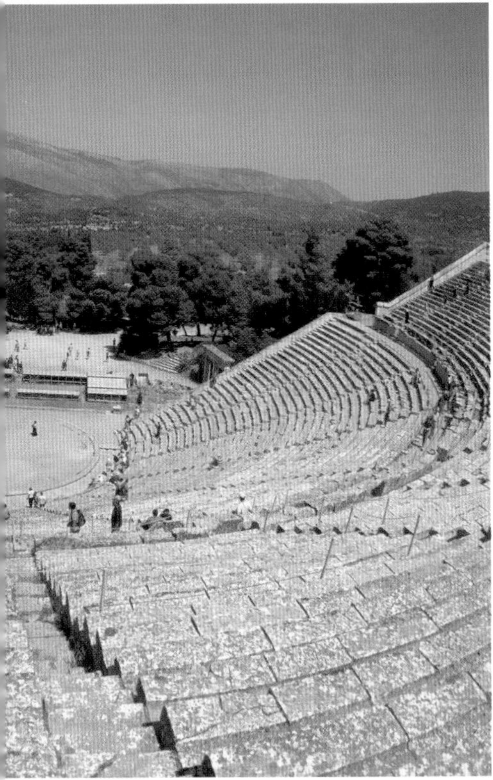

awarded to the best plays presented at the Dionysia and other religious festivals and their authors could become famous.

OPEN-AIR THEATRES

The ancient Greeks staged their plays in huge open-air amphitheatres, which were often sited close to temples and could house as many as 18,000 spectators. They had a semi-circular design, with banks of stone seating built into a natural hillside. The acting area was divided into two, with a large space (the *orchestra*) at ground level in front of a raised stage (the *proscenium*) on which the main actors performed. Behind the raised area was a substantial building (the *skene*), which housed dressing-rooms and other important facilities.

THEATRICAL PERFORMANCES

The earliest theatrical entertainments featured a group of performers called the Chorus, who sang and danced

Acoustics

The semi-circular design of ancient Greek theatres meant they had excellent acoustics (sound properties). An actor speaking on the stage could be heard clearly anywhere in the auditorium, even when whispering.

Audiences

Spectators were admitted to theatres by tickets in the form of bronze discs indicating where a person might sit. Women were rarely, if ever, allowed in. Some theatre staff carried big sticks to strike anyone who tried to start a riot when they disliked a play.

in unison. Around 530 BC, however, a man named Thespis (hence our word 'thespian', meaning actor) introduced a single performer to take leading roles. Later, a second and third actor were added.

These main actors were all male and all professionally trained. They played all the main parts, including women, while the non-professional Chorus were limited to commenting upon the action of the play, singing their lines in unison and dancing.

Masks

All the performers wore masks representing characters they played. Thus they could change from one role to another simply by changing their mask. The masks worn in tragedy were relatively plain, while those worn in comedy were grotesque. The exaggerated features of these masks meant that the spectators could see expressions from a considerable distance. The shape of the masks also made the voices of the actors louder.

As many as four plays might be performed one after the other, the whole entertainment lasting all day. Scene changes were simple, but the special effects could be spectacular, with, for instance, the use of cranes to raise and lower performers playing gods.

The Greek masks used by actors in tragedy and comedy had exaggerated features to aid the audience.

GREEK PLAYWRIGHTS

Greek playwrights were among the world's first great dramatists. Some of their plays are still performed today. Among the most famous writers of tragedy were Aeschylus, Sophocles and Euripides. Aristophanes was the most popular writer of comedy.

AESCHYLUS

Born at Eleusis, near Athens, around 525 BC, Aeschylus is remembered as the Father of Greek tragedy. He was the first playwright to introduce a second actor and to use scenery and costumes. As a soldier, he fought at Marathon and Salamis and some of his plays, such as *The Persians*, reflected his military experience. Most famous of his works is the *Oresteia* trilogy, in which Orestes kills his mother Clytemnestra after she takes a lover, and murders his father, Agamemnon. Orestes is pursued and punished by the Furies, but eventually released when it is decided he has suffered enough. The themes of this and other plays include justice, mercy and wisdom gained through suffering.

Killed by a tortoise

Aeschylus was completely bald. Legend claims that he died (in 456 BC) when an eagle dropped a tortoise on his head to break its shell, thinking his bald head was a rock.

ARISTOPHANES

The most famous writer of comedies for the ancient Greek theatre was Aristophanes. Just eleven of his forty comedies survive, and some are still regularly staged. Most of his prize-winning plays, written in the fourth century BC, poked fun at political events of the day but were also notable for their imaginative plots and poetic language. The most popular included *The Wasps*, *The Birds*, *The Frogs* and *Lysistrata*, in which the women of Athens go on strike until their husbands agree to stop going to war, his message being that the Greek city-states should stop fighting each other and instead cooperate in ruling Greece.

EURIPIDES

The Athenian playwright Euripides was born around 485 BC, probably in Salamis, and won fame as a writer of tragedies. Of the ninety plays he wrote, nineteen survive today. They include *Medea*, *The Bacchae*, *The Women of Troy* and *Orestes*. The plots of his plays, which were notable for their inclusion of strong female characters and clever slaves, typically explore the reasons behind people's actions.

Praise from Sophocles

Euripides later moved to the royal court of Macedonia, where he died in 406 BC. Fellow-playwright Sophocles once said that while he himself showed people as they ought to be, Euripedes showed them as they really are.

SOPHOCLES

The poet and playwright Sophocles was born around 496 BC and wrote 123 plays, of which seven survive. His celebrated tragedies include *Ajax*, *Oedipus Tyrannus*, *Electra* and *Antigone*, and contain much fine poetic language and tackle themes such as right and wrong. Several of his plays concern a central figure who is brought down by a fatal character flaw. Winner of many prizes at the Athens theatre festivals, he developed the use of scenery and was the first playwright to use more than two actors. He died in 406 BC.

 Sophocles' plays are still performed today, as in this modern production of one of his tragedies.

MUSIC AND DANCE

The ancient Greeks loved music and dancing, and both were central to religious ceremonies as well as private and public entertainments. Most children were taught how to dance and play musical instruments, although only females danced as adults.

It was widely believed that music had various magical powers, and that it could inspire love, soothe angry people and tame wild animals. There were songs for most public and private events, including work songs, songs for theatrical entertainments, religious songs, songs sung by soldiers going to war, wedding songs, funeral songs, love songs, drinking songs and songs that could be sung at feasts. We do not know today what the tunes sounded like, however, as very few of them were ever written down.

INSTRUMENTS

The first lyres, which were sacred to Apollo, were made from an empty tortoiseshell and a pair of ox horns, to which strings of animal sinew or hide were attached. Later versions resembled small modern harps. The more elaborate *kithara* was a type of wooden lyre played by professional musicians, who plucked the strings with a plectrum.

The double-pipes (or *auloi*) comprised a pair of pipes with reed mouthpieces. They made a cheerful sound, but were hard to master, with each hand playing a different tune, one of which was a background drone, the other the melody.

Other instruments included the harp, cymbals, the *timpanon* (a tambourine) and the *syrinx* (pan pipes, made with reeds of different lengths).

 This painting shows a Greek couple with a lyre, an instrument that was sacred to the god Apollo.

PHILOSOPHY

The ancient Greeks had a strong philosophical tradition, and the ideas of Greek scholars and thinkers were at the centre of Greek civilization. The ranks of Greek philosophers included thinkers who are still honoured as fathers of modern European culture.

Greek philosophy progressed steadily from the sixth century BC and embraced new thinking in a wide range of fields, including science, biology, mathematics, astronomy, geography, religion and other subjects as well as philosophy as it is understood today.

THE NATURAL PHILOSOPHERS

The first ancient Greek philosophers, who were known as the natural philosophers, contemplated the various ways in which the world works and what things are made of. They rejected mythological stories that sought to explain such natural phenomena as thunder and lightning and instead turned to reason and scientific observation for explanations.

Later philosophers similarly discussed the workings of the universe, but also considered such issues as human behaviour and how society should be governed for the greater good.

SYMPOSIA

The ancient Greek philosophers often developed
and explained their ideas in conversation with other
scholars and thinkers. Some founded schools where
such discussion could take place between philosophers
and the groups of pupils who came to listen to their
opinions. Others met in the *agora* and exchanged ideas
there. Debates on philosophy and other subjects might
also take place between guests (usually all male) at
banquets or at drinking parties and other gatherings
that were called *symposia*.

GREEK PHILOSOPHERS

The first of the great Greek philosophers was Thales of
Miletus who, like others in the pre-Socratic tradition of
natural philosophers, inherited the ideas of Babylonian
and Egyptian scholars but developed them with
greater vigour and imagination. Among the generations
of Greek philosophers who followed his lead, many
names have remained prominent among the world's
great thinkers through the centuries.

Lovers of wisdom

The word 'philosophy' is itself Greek in origin. It means
'lover of wisdom'.

ARISTOTLE

The Athenian philosopher Aristotle was born in the Greek colony of Stagira around 384 BC and became a pupil of Plato. He developed ideas about life and nature, explored politics and science and also devised a method of thinking that he called logic. He discussed his theories at the Lyceum, the school he founded in Athens after returning there from exile in 335 BC.

Aristotle's pupils included the young Alexander the Great, who studied under him for three years. After Alexander's death, Aristotle was forced to flee to Chalcis in Euboea, where he died in 322 BC. His writings, which covered ethics, physics, poetry, psychology and philosophy among many other subjects, included the *Nicomachean Ethics*, *Poetics*, *Politics* and *Metaphysics*.

DIOGENES

Diogenes was a fourth-century BC philosopher from Sinope in Asia Minor, who became the founder of the Cynics, a school of philosophy that rejected the rules by which normal society was governed. He believed in living simply, and condemned all kinds of dishonesty as well as the possession of great wealth.

To demonstrate his beliefs Diogenes took to living in a large barrel or storage jar, thus proving to the world that he was above material things and interested only in ideas. He was also said to wander the streets of Athens with a lamp, looking for an honest man. He died in Corinth, where he had settled after being captured by pirates on a sea voyage.

EPICURUS

Born on the island of Samos around 341 BC, Epicurus founded a school of philosophy in Athens that was to provide a model for many later academic (Epicurean) communities. He argued that happiness was the only proper pursuit of humanity, ideally achieved through a life of peace and simplicity. He doubted that the gods intervened directly in the lives of mortals and rejected the notion that the soul survives the death of the body.

PLATO

Plato was an Athenian philosopher who was born into an aristocratic family around 427 BC and became the most famous of Socrates' pupils. After a time in Syracuse, he went on to found the Academy, a famous school of philosophy set in a pleasant grove in Athens where he passed on his own ideas as well as those of Socrates to Aristotle among other pupils. Here, Plato encouraged open discussion on a range of topics, from the origins of the universe and man's place in the natural world, to the laws of geometry and the best way to govern a country.

It was through Plato that the theories of Socrates were recorded for posterity, usually in the form of conversations. His most famous books included *The Republic*, *Dialogues* and *The Apology*, which was his response to critics of Socrates. He died in 347 BC, but his Academy continued to operate until it was closed on the orders of the Romans in 529 AD.

The School of Plato was the renowned Academy in Athens, where philosophy was studied and discussed.

SOCRATES

Socrates was born into a wealthy Athenian family around 469 BC and served in the army before going on to establish a reputation as a great philosopher, whose views were firmly based on logic and reason. He challenged accepted notions in order to reveal new truths about human behaviour and good and evil, arguing that it was natural for people to behave well, as long as they knew what good behaviour was. He also debated the nature of beauty and happiness, among other subjects.

Radical ideas

Although he had a strong sense of humour, Socrates could be highly critical of those who disagreed with him, earning himself the nickname the Gadfly, because his arguments 'stung' those who opposed him. His favourite method of defeating his opponents in argument was to trick them into contradicting

Modest genius

Socrates wrote down nothing, exploring his ideas through discussion with others, but his conclusions were recorded by his pupil Plato. A pot-bellied, ugly man, he was kindly and modest, and once observed: 'The only thing I know is that I know nothing.'

themselves (a 'Socratic argument'). His radical ideas made him very controversial, however, and in 399 BC, after being accused of corrupting the young and offending the gods, he was forced to commit suicide by drinking hemlock.

THE STOICS

The Stoics was an influential school of philosophers that first met around 300 BC in the Stoa Poikile (Painted Porch) in Athens. Under their leader Zeno of Citium, the Stoics discussed the basis of the universe and stressed the importance of living in harmony with nature. The doctrine of stoicism recommended a life of calm and reason guided by virtue and duty.

GRECO-ROMAN PHILOSOPHY

During the Hellenistic era, when Greece finally became part of the Roman world, Greek ideas of many kinds were absorbed into Roman culture, and the theories of the Greek philosophers were similarly translated and further developed. The ideas of Plato, for instance, inspired the Neo-Platonism of Plotinus and other Roman thinkers, while the Stoicism of Zeno of Citium influenced such Roman philosophers as Seneca and Epictetus. Further philosophical speculation was ultimately slowed, however, by the rise of Christianity during this period.

HISTORY

The ancient Greeks are often said to have produced the world's first historians. Their earliest historical studies date from the sixth century BC, when the need to understand their enemies the Persians prompted them to research Persian culture and history. The word 'history' itself comes from the Greek word for 'enquiry'.

HERODOTUS

Born in the Greek colony of Halicarnassus in Ionia around 485 BC, Herodotus is often called the Father of History. He wrote the first prose accounts of current affairs, concentrating upon the wars with Persia but also making records of the history of Lydia, Babylon and Egypt. His history of the Persian Wars was largely based on interviews with those involved. He gathered much of his material while travelling, visiting Egypt, the Black Sea regions, Babylon and Cyrene, and living on the island of Samos, in Athens and in southern Italy.

Accuracy

Opinions about the accuracy of the nine volumes of history written by Herodotus have varied over the centuries. Some material appears to have been speculation, but much else has been supported by archaeological evidence.

THUCYDIDES

The Athenian historian Thucydides was born near Athens around 460 BC and commanded an Athenian fleet in the Peloponnesian War between Athens and Sparta. After losing his command, he wrote an eight-volume history of the Peloponnesian War, based on his personal experience. He is said to have been assassinated after returning to Athens in 404 BC.

TIMAEUS

Born at Tauromenium in Sicily around 345 BC, Timaeus was forced to leave the island by Agathocles, of whom he had been critical, and spent the next 50 years in Athens. He is remembered for his forty-volume *Histories*, which covered the history of Greece up to the first Punic war. He also foresaw the rise of Rome. Later admirers of his writing included Cicero and Plutarch.

XENOPHON

Born in Attica around 430 BC, Xenophon was a pupil of Socrates who went on to fight as a mercenary general in several campaigns. Banished from Athens for siding with Sparta, he went into retirement and devoted himself to writing historical records of the Persian Wars, as well as a history of Greece and works on military tactics, politics and horsemanship.

SCIENCE AND MATHEMATICS

The ancient Greek scholars studied all aspects of science and mathematics, although they made little distinction between the different fields. Building upon the significant advances in scientific and mathematical understanding made by the ancient Egyptians, the Greeks went on to make numerous discoveries and they established the foundations upon which modern science and mathematics rest. Particularly important were their ideas about the relationship between scientific and mathematical theories and on the nature of the physical world around them.

NOTABLE DISCOVERIES

Greek astronomers not only predicted solar eclipses but also discovered that the Earth was round and floats freely in space, turning on an imaginary line called an axis. Other scientists found out more about time and the seasons through their study of the stars, worked out the circumference of the Earth, explored geometry and algebra, theorized about human evolution, advanced their understanding of levers and pulleys and developed theories about atomic structure. The Greek geographers made maps of both the Earth and the sky and devised the lines of latitude and longitude that are still used today.

Mistakes

Greek scientists and mathematicians were not always right. They were wrong to assume, for instance, that the Earth is at the centre of the universe. Their invention of numerology (the study of the magical properties of numbers) gave birth to astrology, meanwhile, but is otherwise of little interest to modern scientists.

The Greeks were clever at applying their scientific and mathematical discoveries to practical ends, inventing new tools and weapons as well as scientific instruments that made further discoveries possible.

GREEK SCIENTISTS AND MATHEMATICIANS

The theories of ancient Greek philosophical thinkers laid the basis for modern scientific and mathematical disciplines. Their important contributions to such fields as geometry, number theory, applied mathematics, astronomy, medicine and the theory of matter changed irrevocably the way in which subsequent generations of scholars saw the world.

ANAXAGORAS

Born in Clazomenae in Asia Minor around 500 BC, Anaxagoras was a scientific philosopher whose work was to have a profound influence upon other scientists.

He explained solar eclipses and discovered that the moon did not produce light itself but merely reflected the light of the sun, which he concluded was a ball of fiery material. In *On Nature*, he tried to explain further the workings of the universe, arguing that all matter could be divided endlessly into smaller particles. His pupils included Euripides and the statesman Pericles.

ANAXIMANDER

Born in Miletus, Anaximander was a pupil of Thales who made significant advances in the understanding of time, astronomy and the evolution of life on Earth. Sometimes called the Father of Astronomy, he measured the length of the seasons, made what is said to be the first map of the known world and realized that much of the land had once been covered by water. He also developed theories about the early evolution of humans, suggesting that they had evolved from more primitive creatures that had emerged from the sea.

ARCHIMEDES

Born in Syracuse around 287 BC, Archimedes was the son of the astronomer Phidias and won lasting fame himself as a mathematician, astronomer and inventor, living for many years at the royal court in Syracuse. Particularly significant were the mathematical formulae he devised to calculate the area and volume of spheres, cylinders and other figures. He also predicted eclipses, worked out distances to the stars, measured the year and applied the principle of levers and pulleys to move large objects (once moving a ship over dry land using just one hand and a pulley). Equally practical was his invention of the Archimedes screw, which enabled water to be raised from one level to another.

MILITARY INVENTIONS

Other inventions had military uses, among them huge catapults and cranes with mechanical claws that could lift enemy warships out of the water. Such was his reputation that the Romans are said to have fled at the mere sight of his war machines. When Syracuse was attacked by a Roman fleet, he had the soldiers polish their bronze shields and stand on the quayside in a huge curve, forming a massive mirror, to deflect the sun's rays towards the Roman ships and set fire to them.

THE EUREKA MOMENT

The most famous episode in his life occurred when Archimedes was challenged to prove that the royal jeweller at Syracuse had made the king's crown with less gold than he claimed. When Archimedes, pondering the problem, took his bath, the water overflowed, demonstrating that his body displaced a volume of water equivalent to its own volume. The great man ran naked into the street shouting 'Eureka!' ('I have it!'), realizing that if he placed the crown and a piece of gold of the same weight in a bowl of water both should displace the same amount of fluid.

The death of Archimedes

Archimedes died in 212 BC, apparently killed by a Roman soldier in Syracuse after he failed to respond, when challenged, until he had finished his calculation.

ERATOSTHENES

Born in Cyrene around 276 BC, Eratosthenes became head of the famous library at Alexandria in Egypt and was respected for his extensive knowledge in a wide range of fields, from mathematics and mechanics to geography and literature.

He is especially remembered for his surprisingly accurate calculation of the circumference of the Earth, which he worked out by recording the angle of the Sun at various different places.

EUCLID

Euclid established a school of mathematics in Alexandria around 300 BC and made several major discoveries in mathematics and geometry. His *Elements* summarized the mathematical knowledge of his day and remained a standard text in schools into the twentieth century. He also wrote on astronomy, music and optics.

Overleaf: A mosaic depicting the death of the Greek mathematician and inventor Archimedes.

PYTHAGORAS

Probably born on the island of Samos around 580 BC, Pythagoras was a philosopher and mathematician whose ideas were to have a profound influence on future generations. His life is shrouded in legend, but it appears that he set up a school in the Greek colony of Croton in southern Italy that developed into a 'Pythagorean' community following a way of life guided by his religious and scientific ideas.

Heralded as 'the most noble philosopher among the Greeks' by Herodotus, Pythagoras believed the universe was governed by mathematical patterns and that mathematics and religion were closely related. He also believed in reincarnation and that the souls of the dead were reborn in other bodies depending on how well they behaved in life. He and his followers studied the properties of shapes and angles, fractions, odd and even numbers, musical intervals and the movements of the stars.

INFLUENCE

Pythagoras' Theorem on right-angled triangles and his other discoveries are still taught today (although he was wrong to think the Sun went round the Earth). He died around 500 BC, but his influence continues to be felt. He was much admired, for instance, by Albert Einstein, who further developed his theories in the twentieth century.

THALES

Born around 620 BC, Thales's interests ranged from mathematics and engineering to politics and astronomy. The founder of the Greek philosophical tradition, Thales visited Egypt and conveyed their mathematical knowledge to Greece. He is said to have predicted a solar eclipse in 585 BC, to have worked out the height of a pyramid by measuring its shadow and to have calculated the number of days in a year and the length of the seasons. Less accurate were his notions that the Earth was flat and floated on water.

THEORIES OF MATTER

One of the issues that most preoccupied ancient Greek scientists and mathematicians was the nature of matter. According to Thales, all things came originally from water. Anaximenes, however, believed that air was the basic substance of matter. Heraclitus of Ephesus emphasized the element of fire in changing the nature of substances, while Empedocles of Acragas combined these ideas and argued that there were four basic elements: earth, water, air and fire. Leucippus and Democritus developed the theory that matter could be subdivided into atoms, which could not be broken down further. Plato, building on Pythagoras, stressed that material things were reflections of eternal, unchanging ideas and focused on the underlying phenomena that led to the creation of matter.

MEDICINE

The Greeks inherited the ancient Egyptians' interest in medicine and greatly advanced understanding of medical matters. Although superstitions and traditional folk remedies were still important, it was gradually realized that good human health depended largely on following a healthy lifestyle, based on such factors as a sensible diet, exercise, adequate sleep, good sanitation and clean water.

Good hygiene

To prevent the spread of disease, the Greeks kept their water supply separate from their drainage system, and washed themselves frequently with olive oil, which they scraped off with a tool called a strigil. Homes and clothes were also kept as clean as possible. Such measures seem to have been effective, as some men and women lived into their seventies or beyond.

Medicines and procedures

The sick were treated with medicines that were made from plant extracts or, if necessary, they were subjected to an operation. Archaeologists have found various medical tools, including saws to cut off limbs, cups to catch blood and tweezers to extract spearheads. Medical doctors were highly respected and well-paid members of Greek society.

 A fanciful relief depicting Asclepius, the god of healing and medicine, healing a patient.

FAITH HEALING

Many ancient Greeks believed that illness was sent by the gods and that the only cure was to regain their favour. Sick people were encouraged to sleep near temples that were dedicated to Asclepius, the god of healing and medicine. Prayers and offerings would also be presented to the gods and, if cured, a patient might hang a model of the repaired limb or organ on the wall of the temple in thanks.

Operating theatres

Some patients were recommended to visit the theatre to watch a sacred drama being performed, believing that this might magically cure them of their ills.

HIPPOCRATES

Called the Father of Medicine, Hippocrates founded a school of medicine on the Greek island of Cos, where he had been born around 460 BC. Here he passed his ideas on to his pupils, who gathered under a plane tree to hear his opinions.

Hippocrates recommended treating the human body as a single organism and basing diagnoses on careful examination and questioning of the patient, preferring a scientific approach to reliance upon magic or religion. He treated disease through natural remedies, though he also favoured bloodletting (a common practice for centuries afterwards). He was expert in treating broken bones and correctly concluded that air is carried in the veins. He also stressed the importance of high ethical standards and winning the trust of patients.

The four humours

Some of Hippocrates' ideas have been proved wrong over the centuries, but they were highly influential in their time. These included the notion that the body contains four humours (blood, phlegm, yellow bile and black bile) and that it is an imbalance between these that causes illness. His belief that epilepsy was caused by too much phlegm and too little air in the brain was also mistaken.

The death of Hippocrates

Hippocrates died around 377 BC. He left 53 books, including *On Sacred Disease*, which was about epilepsy, *Wounds in the Head* and *Women's Diseases*.

The Hippocratic oath

The so-called Hippocratic oath was taken by new doctors well into the twentieth century. In this they agreed not to harm patients and to keep what they learned about their patients' conditions confidential.

PART 6

War in ancient Greece

The city-states of ancient Greece often fell out with one another and frequently settled their disputes on the battlefield. Athens and Sparta in particular fought each other many times, and their warring eventually brought Greek civilization to its knees. Occasionally, however, the Greek states also formed military alliances to defend themselves against enemies from beyond Greek borders, notably the Persians.

Arms and armour

This is a well-preserved example of an ancient Greek helmet.

ARMIES

Amongst the ancient Greek city-states, only the Spartans maintained a full-time army, training boys as soldiers from a very young age and organizing their whole society around preparation for warfare. The other Greek city-states relied instead upon calling upon their fit and able citizens to fight as and when they were needed.

THE ATHENIAN ARMY

The city-state of Athens arranged two years of military training for young men at the age of eighteen and formed its armies from men aged between twenty and fifty years, while those older men who were aged between fifty and sixty were kept in reserve or ordered to man the garrisons. Most Athenian citizens obeyed such summonses, wishing to protect their homes and families from potential enemies.

Spartan heroes

Spartan warriors were among the most formidable. They wore their hair long (carefully combing it before battle). If they proved to be cowards, they had half their hair and beard shaved off – a great disgrace.

This painting shows a Hoplite warrior, armed with a shield and spear.

HOPLITES

Most ordinary citizens served as foot soldiers or in auxiliary units of archers and stone-slingers. In the beginning, most men came from the lower classes and were expected to provide their own equipment, hence they usually carried weapons and armour that were of relatively modest quality. Later, however, as the city-states grew richer, the middle classes were able to provide themselves with better equipment and joined the ranks of the hoplites (spear-carrying foot soldiers equipped with helmets, shields and body armour).

CAVALRY AND COMMANDERS

The wealthiest citizens generally served in the cavalry, providing their own horses and weapons. The army of Athens was led by ten commanders (*strategoi*), who were elected by the Assembly and represented the ten tribes that made up Athenian society.

WEAPONS AND ARMOUR

The soldiers in most Greek armies had to provide their own weapons and armour. There were no standard uniforms, although Spartan warriors usually wore scarlet and the Athenians carried shields which were decorated with the letter 'A'.

POOR SOLDIERS

The poorest soldiers might have little more than animal skins and a wooden club or slingshot. Others might wear armour made from layers of stiffened linen (which was cooler than metal but less effective) or a cuirass made of leather and fight with short swords or bows.

WEALTHY SOLDIERS

Wealthier soldiers serving in the ranks of the hoplites wore a bronze breastplate and helmet and carried a large round shield (called a hoplon) and a spear

Thanks to the gods

If victorious in battle, a warrior might present his armour to the gods, leaving it in a temple or hanging it on the branches of a tree.

or sword. Their lower legs were protected by bronze shin guards that were called greaves. Many Greek warriors attached coloured horsehair plumes to their helmets to make them appear more impressive to their enemies. The richest noblemen fought on horseback or from chariots, carrying a variety of spears, javelins and swords.

SIEGE WEAPONS

When they were involved in laying siege to a town or a city, the ancient Greeks had a number of siege weapons from which to choose. These included catapults, siege towers and battering rams, as well as flame-throwers, which sprayed fire from a cauldron of burning coals and sulphur onto the wooden walls of the enemy's defences.

An ancient Greek helmet. Many helmets were embellished with some colourful horsehair plumes.

TACTICS

The early battles in the ancient Greek world were dominated by cavalry, but by the time of the Archaic Period the outcome of most battles was decided by action between foot soldiers.

THE PHALANX

The hoplites of ancient Greece (see page 211) fought shoulder to shoulder in packed formations that were called phalanxes. Each man held his shield so that it overlapped that of his neighbour to the left, meaning that the enemy was confronted with a formidable unbroken wall of shields.

Such formations were effective only so long as each man trusted the man fighting next to him. Each phalanx was at least six men deep, so that men in the rear ranks could fill any gaps left by dead or wounded men in front of them.

Defensive formations

The phalanx was essentially a defensive formation, so commanders preferred to place their troops in a strong defensive position and wait until attacked. If phalanxes fought each other both would try to outflank their opponents' right wing, which was less well-protected.

ATTACKING AN ENEMY

When attacking, the Greeks made great use of cavalry and chariots to throw enemy formations into disarray. The Thracians, meanwhile, developed a tactic in which they used javelin-hurling soldiers called peltasts to break up phalanxes and pick off individual warriors. To counter the peltasts, the Greeks would order some of their fittest men to remove their armour and rush out from the phalanx to chase away such opponents.

Cavalry actions

With the introduction of large phalanxes of hoplites, the cavalry generally played a supporting role in battle. They were usually restricted to exploiting weaknesses in the enemy formation created by the foot soldiers and to pursuing disordered foes. The Athenians put just 300 horsemen into the field during the Persian wars, but the usefulness of mounted men as scouts and as a means to break up enemy formations persuaded them to increase their number to 1,000 by the fifth century BC.

SIEGES

Many cities (like Troy) fell after long sieges, which could last for years. When laying siege to a town that could not be taken by direct assault, attacking troops burned crops in surrounding fields and cut off supply routes, so that the defenders would be starved into surrender.

NAVIES

Most of the Greek city-states became rich through maritime trade, and they built up substantial fleets of warships to protect this trade and to fend off enemy fleets carrying invading armies. Greek fleets routinely patrolled the Aegean Sea and other areas of the eastern Mediterranean.

WARSHIPS

The Minoans of Crete improved early ship design around 1600 BC, making longer voyages at sea possible for the first time. The high prows of the Minoan galleys helped the ships cut through the water, while a large oar, which was used as a rudder, allowed them to be steered more accurately.

Sail and oar power

The warships of the time combined sail and oar power. A flute-player played tunes to maintain a steady rhythm among the oarsmen. The largest and fastest warships were the triremes, which first appeared in the sixth century BC and carried up to 170 oarsmen who were mounted in three banks on each side of the vessel, with one man to each of the twenty-five oars in each row. Athens, at the height of its power and influence, had a fleet of 300 triremes.

Lucky eyes

Many ships had a big eye painted on each side of the prow. It was believed that these would scare away evil spirits and protect the crew.

Weapons on board

Warships were equipped with a bronze ram (a sharp spike) that could be driven into the planks of an enemy ship either to sink it or make it vulnerable to boarding. The largest ships also carried catapults and ballistas for use in sieges or to throw burning missiles of naphtha, sulphur and saltpetre ('Greek Fire') onto enemy ships and set fire to them.

CREW

The crew of a typical trireme of the Classical Period numbered 200, with five officers. The captain (*trierarchos*) was assisted by an executive officer (*kybernetes*), an officer responsible for training and morale (*keleustes*), an administrative officer (*pentecontarchos*) and an officer responsible for keeping a lookout (*procrates*). The rest of the crew comprised the musician who kept the oarsmen in rhythm (*auletes*), 170 oarsmen, fourteen marines (ten of whom were armed with spears and four with bows and arrows) and ten sailors who were responsible for handling the sails.

MAJOR WARS

The history of ancient Greece included numerous wars between the city-states and against invading armies. There was also conflict with the many pirates who preyed upon the merchant ships sailing through the eastern Mediterranean.

Athens and Sparta were among the most powerful city-states and they fought repeatedly to gain control over the region. Conflict between the two states was at its most intense during the Peloponnesian Wars of the fifth century BC. Sparta's well-trained professional army ultimately defeated the volunteer army of the Athenians, but in the Corinthian War that followed in the years 395–387 BC, it was confronted by the allied forces of Corinth, Athens, Argos and Thebes and was eventually beaten by the Theban army at Leuctra.

INVASIONS

Many of the most famous battles involving Greek warriors were fought in the course of wars against invading armies from beyond Greek borders. Notable among these were the long-running Persian Wars that began towards the end of the sixth century BC and featured such classic engagements as the battles of Marathon and Thermopylae. Greek defeat against the invading armies of Philip II of Macedon in 338 BC, meanwhile, spelt the end of Greek independence. After that date, Greek soldiers were recruited to serve in the occupying armies of Macedon and, eventually, Rome.

MYTHICAL WARS

Greek legend also retold the adventures of heroes fighting against the Amazons – a mythical race of women warriors from north of the Black Sea, possibly inspired by the female warriors who were included in the Scythian army – and against the armies of Troy in the Trojan War, among other formidable opponents who may or may not have existed.

THE PERSIAN WARS

In the sixth century BC, the Persians, from modern Iran, conquered Greek territories in Ionia in western Asia Minor and then threatened to add mainland Greece itself to their growing empire. Over a period of twenty years, the Greeks won a series of remarkable victories against superior Persian forces.

The Battle of Marathon

One of the greatest Greek victories of all took place in 490 BC, after perhaps 60,000 Persians came ashore near Athens. Athens and its ally Plataea quickly raised 11,000 men, but help from Sparta was not forthcoming.

The unequal armies faced each other across the plain of Marathon. The Greek general Miltiades knew that the Persians always put their best men in the middle of their line, so placed his own best soldiers on the flanks and surprised the Persians by attacking first. The Greeks broke the Persian flanks and then attacked the now exposed Persian centre from behind. At the end of the battle, 6,400 Persians lay dead, at a cost of just 203 Greek casualties.

The Battle of Thermopylae

In 480 BC, the Persian and Greek armies met in the narrow mountain pass of Thermopylae in northwest Greece. A small fighting force of 300 Spartans and 700 Boeotians under King Leonidas heroically held the pass against the much larger Persian army of King Xerxes, until finally a Greek traitor named Ephialtes led the Persians around behind their opponents. Leonidas and his few remaining troops were massacred to a man. The heroic sacrifice made by King Leonidas and his men became an enduring legend and is remembered as one of the greatest last stands in military history.

The Battle of Salamis

Later in 480 BC, the Greek and Persian fleets met in a decisive naval battle off the island of Salamis near Athens. The Greeks, under Themistocles, lured the Persian fleet into narrow straits – where they had little room for manoeuvre – and then shattered their forces.

THE PELOPONNESIAN WARS

In 431 BC, the city-state of Corinth fell out with its colony on the island of Corcyra (modern Corfu), resulting in a battle, and with Athens and Sparta (which feared the growing power of the Athenians) getting involved on opposite sides. Sparta and its various allies in the Peloponnese region of Greece supported Corinth, while Athens and the city-states of the Delian League took up the cause of Corcyra.

Stalemate

Over the next 27 years, Sparta proved to be the more powerful force on land, but Athens ruled the seas with its superior navy, so that neither side was able to claim a final victory. Athens itself – protected by the Long Walls built to connect it to the nearby port of Piraeus – successfully resisted a siege by the Spartan army.

Betrayal

A peace treaty was signed between the two sides in 421 BC, but conflict soon flared up once again. An Athenian attack on Syracuse in 415 BC went disastrously wrong after the Athenian commander Alcibiades changed sides and helped the Spartans destroy the attacking force, most of whom were killed or taken prisoner. The Spartans then built a fleet with which they destroyed the Athenian navy at anchor at the Battle of Aegospotami in 405 BC.

The Battle of Leuctra

The victory of the Thebans at Leuctra in Boeotia in 371 BC ended Spartan dominance. Under King Epaminondas, the smaller Theban army won the battle through a change in tactics, shattering the Spartan line by attacking it with a massed column in staggered formation.

Athens surrendered after a renewed siege, and in 404 BC the wars ended in Spartan victory.

Hollow victory

The Spartan victory in the wars changed the balance of power in ancient Greece. The Delian League broke up, and Athens was subjected to the dominance of Sparta. However, the costs of war had severely damaged the economy of the region and, while Athens never again returned to its former greatness, the whole of the Peloponnese was reduced to a state of poverty. New civil wars, marked out by a ferocity hitherto unknown in Greece, broke out between the city-states.

The Spartans made a treaty with their old enemy Persia and then became embroiled in war with Thebes, now allied with Athens, and were disastrously defeated at Leuctra in 371 BC. The Thebans then called on Philip II of Macedon for help. He responded by extending his kingdom to include all of Greece.

ALEXANDER THE GREAT

Most famous of all the warrior leaders of ancient Greece was Alexander the Great. He was the son of King Philip II of Macedon, who had conquered most of the Greek city-states in 338 BC and had thus become the first king to rule the whole of Greece.

CRUSHING VICTORIES

Born around 356 BC, Alexander became king of Macedon at the age of twenty on his father's murder in 336 BC. Mounted on his stallion Bucephalus, he led his army of 35,000 men eastwards to some crushing victories over the Persians.

Over the next eleven years, Alexander conquered much of the known world, including Persia and most of Asia Minor, even going as far east as the Ganges in India. When he tried to go further, however, his men rebelled, and, after sulking in his tent for several days, Alexander reluctantly agreed to turn back.

LOVE OF KNOWLEDGE

Alexander was brave, intelligent and popular with his men. He could also be violent, however, and he killed his foster-brother Kleitus in a drunken brawl and once burned down a palace during a wild party. His adoption of eastern habits led some to accuse him of not being truly Greek, but his conquests greatly extended Greek influence and promoted the fusion of Greek culture with that of the Middle East during what became known as the Hellenistic Age.

Having been a pupil of Aristotle, Alexander had a great love of knowledge and was said to carry copies of Homer's *Iliad* on campaign. The city of Alexandria in Egypt, one of seventy cities he is said to have founded, remained a centre of scholarship for centuries.

PREMATURE DEATH

Alexander's premature death of a fever in Babylon in 323 BC deprived the Greek world of its greatest military leader and without him his empire quickly disintegrated.

> ### Barbarian Greeks
>
> Macedon, in northeast Greece, was an unlikely place to produce one of the great heroes of the Greek world. Most Greeks struggled to understand the thick accent of Macedonians and considered them barbarians.

Opposite: A sculpture of Alexander the Great, who extended Greek influence throughout the ancient world.

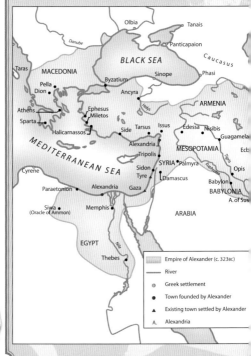

The empire of Alexander the Great

Olbia · Tanais

Danube

BLACK SEA

Caucasus

Taras · MACEDONIA

Byzantium · Sinope · Phasi

Pella · Dion ·

Athens · Ephesus · Ancyra · ARMENIA

Sparta · Miletos

Halicarnassos · Side · Tarsus · Issus · Edessa · Nisibis · Guagamela

MEDITERRANEAN SEA · Alexandria · MESOPOTAMIA · Ecb·

Tripolis · SYRIA · Palmyra · Opis

Cyrene · Sidon · Damascus · Babylon

Tyre · BABYLONIA

Paraetonion · Alexandria · Gaza · A. of Sus·

Siwa · ARABIA
(Oracle of Ammon) · Memphis ·

Nile

EGYPT

Thebes

	Empire of Alexander (c. 323 BC)
	River
◎	Greek settlement
●	Town founded by Alexander
▲	Existing town settled by Alexander
A.	Alexandria

PART 7

Greece today

Visitors to modern Greece can still see many impressive ruins that date back to ancient times, and the museums of the world house thousands of fascinating finds which have been recovered by generations of archaeologists. But the legacy of ancient Greece is also evident in many aspects of modern life, from art and science to philosophy and language.

Ruined temples
The ruins of the Temple of Poseidon, which overlook the Aegean Sea, are still visited today.

REDISCOVERING GREECE

Historians have found out much about the lives of the ancient Greeks from various sources. Many of the ideas of the Greek philosophers, statesmen and writers have been passed down to us through the Roman and later civilizations, and are now enshrined in modern Western society in fields ranging from literature and science to theatre and the law.

The modern rediscovery of ancient Greek civilization dates back largely to the first half of the nineteenth century. The sites of ancient Greece had remained mostly ignored (often half-buried) between the end of the Byzantine Empire in 1453 and the arrival of the first wave of European travellers around the beginning of the nineteenth century. There were precious few descriptions of ancient sites and even fewer visual records of the ruins that remained.

Archaeological finds

Archaeologists have located and examined many important ancient Greek sites, and thousands of their finds may be seen in museums around the world. These include numerous notable finds recovered from the seabed by marine archaeologists.

As Turkish authority weakened, access to the area became easier. The outbreak of the Revolutionary and Napoleonic wars in France, meanwhile, limited travel opportunities in Europe itself and encouraged intrepid travellers to explore further afield. The declaration of independence by the revolutionary Greek state in 1834 proved an open invitation to artists, historians and archaeologists to make the first proper studies of the region and its antiquities.

WRITTEN HISTORY

The Greeks were one of the earliest people to record the history of their civilization in writing. Although the papyrus scrolls on which they wrote mostly perished many hundreds of years ago, they were repeatedly copied by later writers, and thus can still be read today. Over the centuries, scholars have studied classic Greek texts on history, philosophy, politics, medicine, literature and other academic subjects, as well as more practical documents relating to trade and diplomacy between the city-states, among other topics.

The Greek language

The language of the ancient Greeks is no longer spoken, but as a written language it continues to be studied and understood by scholars throughout the world. Many ancient Greek words have been absorbed into modern languages.

GREEK ARCHAEOLOGY

Finds unearthed by archaeologists have ranged from the remains of temples and other important buildings to smaller but equally significant objects, such as statues, ships, pottery, weapons, jewellery and coins.

Because most ordinary Greek houses were made of sun-dried mud bricks, little trace of these has survived. The Spartans, in particular, built virtually nothing out of stone, so very little remains of their towns and villages. More important buildings, however, were usually made of stone, and ruins such as the Parthenon in Athens have stood through the centuries as a testament to ancient Greek culture and achievement.

GREEK ARCHAEOLOGISTS

Many notable scholars built substantial reputations upon their work on Greek archaeology. The most famous have included Sir William Hamilton, a British diplomat and classicist of the early nineteenth century whose collection of vases is now kept in the British Museum; the amateur German archaeologist Heinrich Schliemann, who identified the site of Troy in 1870; and Sir Arthur Evans, the British aristocrat who explored Knossos on Crete in the 1890s and unearthed spectacular evidence of early Minoan civilization.

Potted history

The discovery of thousands of ancient Greek pots and vases over the centuries has proved particularly valuable, as these were often decorated with scenes depicting everyday life in ancient Greece, providing a great deal of information about what ordinary people did. Important discoveries of Greek pots have been made in areas far remote from Greece itself, as items of pottery were traded with many distant peoples, from Spain to the Ukraine.

Revival of the Olympic Games

It is worth mentioning the French aristocrat Pierre de Coubertin, who revived the tradition of the Olympic Games in Athens in 1896. Coubertin had been inspired by the recent archaeological discoveries at Olympia, the home of the original Olympic Games.

Specialists

Given the geographical extent and long history of ancient Greek civilization, archaeologists interested in this period usually specialize in a particular region or age, such as Neolithic Greece, the Minoans of Bronze Age Crete, Troy, the Golden Age of Athens or peoples and events of the Hellenistic era. It is evident from the diversity of these studies that ancient Greece was not a single unified culture but a civilization based on a combination of cultures with regional variations.

WHAT YOU CAN SEE TODAY

The impressive physical remains of ancient Greek civilization may still be seen at several important sites throughout Greece and its colonies. The most famous is unquestionably the Acropolis in Athens, which is still crowned by the ruins of the Parthenon temple and traces of associated buildings. Over the centuries, the Parthenon has been used as a Christian church, a mosque and even as an arsenal by invading Turks, until it was blown up by the Venetians in 1687.

IMPORTANT SITES

Other important sites with substantial ruins, which can still be visited today, include the Oracle at Delphi, which includes the remains of the Sanctuary of Apollo and the Castalian Spring (where visitors bathed), the palace of Knossos on Crete, the palace and royal graves of Mycenae, the theatres at Dodona and Epidauros and the stadium at Olympia, as well as other temples and theatres elsewhere in Greece, Turkey and Sicily.

TROY

The site of Troy, the scene of the epic events described in Homer's *Iliad*, was identified in 1871 by the German archaeologist Heinrich Schliemann. He located the site

German archaeologist Heinrich Schliemann depicted sketching the ruins he identified as those of Troy.

at Hisarlik, in modern Turkey, and later archaeologists have unearthed traces of a substantial city here, which was apparently founded around 3600 BC. It remains to be proved beyond doubt, however, that the events described by Homer actually took place.

Overleaf: The ruins of ancient Corinth, formerly second only to Athens among the city-states.

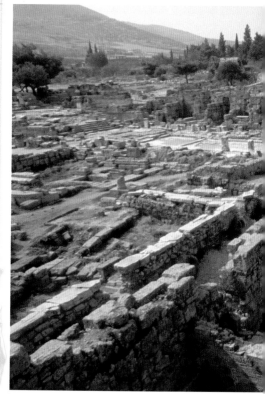

THE CULTURAL LEGACY

More important than the physical remains of the ancient Greek civilization was its cultural legacy. The conquest of Greece by the Roman legions in the first century BC greatly increased contact between the two cultures. The Romans had a deep admiration for many aspects of ancient Greek civilization and they often chose educated Greek slaves to act as tutors to their children. The Romans (who also absorbed many features of Germanic, Slavic and Celtic culture) did not take up all the ideas developed by the Greeks, however, and it was not, for instance, until many centuries later that other countries attempted to revive the ancient Greek ideal of democratic government.

CLASSICS

Greek thinkers, such as Socrates, Plato and Aristotle, laid the foundations of modern Western philosophy, and

The sporting legacy

The legacy of the ancient Greeks even extended as far as the world of sports. The modern Olympics, first staged in 1896 on the model established by the ancient Greeks, remain today the top event in world sport.

established many of the fundamental concepts behind modern mathematics and science. The myths related in Greek literature and theatre became central classics in the European literary and dramatic tradition, while such political innovations as the invention of democracy have had a profound influence upon world history.

Architecture

The most visible evidence of the influence of ancient Greek culture over the centuries may be seen in the history of architecture. The ideals of ancient Greek architects, concerning harmony and proportion, reappeared in Europe during the Renaissance of the seventeenth century and have remained important ever since, especially in the so-called neoclassical style, with its symmetrical proportions and pillared porticos and colonnades.

A CLASSICAL EDUCATION

The study of the classics, comprising examination of the language, literature, history and art of ancient Greece and Rome, became central to European school and university education in the nineteenth century, when such knowledge came to be considered a measure of a person's cultural status. Such studies are now included as a branch of learning known as the humanities. Students of the subject are variously identified as humanists or classicists.

LANGUAGE

The extent to which ancient Greek culture became absorbed into that of the modern world may be clearly seen in the many words of ancient Greek origin that are still present today in the vocabulary and common usage of modern European languages. These extend to virtually every field of human activity, from science and mathematics to philosophy, the arts and education. Many of these words were not actually used by the ancient Greeks, but have been constructed over the centuries, as needed, by combining shorter Greek words with relevant meanings.

Modern Greek

The Greek language spoken in Greece today is directly descended from that of the ancient Greeks. With a documented history of around 3,500 years, it is the oldest of all the Indo-European languages. During the Hellenistic period, various Greek dialects merged to form a common tongue, which spread among Greek communities throughout the Mediterranean area. The version of Greek spoken by the armies of Alexander the Great was taken up by the Romans, and in due course Greek established itself as a second language across the Roman empire, after Latin. The use of Greek as the language of the New Testament spread knowledge of the language throughout the Christian world. Greek was also the official language of the Byzantine empire.

Words of Greek origin

The following list comprises a selection of words of ultimately Greek origin that have been absorbed into the vocabulary of the English language.

WORD	MEANING
academy	from Akademia (Plato's grove)
alphabet	from *alpha* and *beta* (letters)
amnesia	without memory
anonymous	without name
anthropology	the study of man
aristocrat	best ruler
astronomy	study of the stars
attic	of or relating to Attica
auto-	self-
bio-	life-
canopy	mosquito net
cathedral	seat
catholic	universal
cemetery	dormitory
chronology	the study of time
cosmos	all, everyone
crypto-	hidden-
cynic	dog
democracy	government by the people
dinosaur	terrible lizard
drama	something performed
economy	domestic management
ego-	self-
epitaph	over a tomb
genesis	beginning

hemisphere	half a sphere
holocaust	entirely burnt
kilo-	thousand-
-logy	study of
martyr	witness
mathematics	something learnt
mega-	big
metropolis	mother city
micro-	small
museum	home of the muses
myriad	ten thousand
mystery	secret rites
palaeo-	old
panic	terror induced by Pan
panorama	full view
philosopher	lover of wisdom
phobia	fear
poly-	many-
psyche	spirit, soul
pylon	gateway
sarcasm	rending the flesh
scene	tent, stage
skeleton	dried up
stigma	brand
stoical	impassive
symphony	same sound
tele-	far
therm-	heat
thesaurus	treasure
thespian	of, or relating to, the theatre
toxic	poisoned arrow
tripod	three feet
xenophobia	fear of strangers

MUSEUMS

Many significant archaeological finds from ancient Greece may now be seen in museums located on different continents around the world.

MAJOR COLLECTIONS IN GREECE

The most important museum of ancient Greek antiquities in Greece is the National Archaeological Museum in Athens. Opened in 1891, it houses many great treasures, such as the Marathon Boy bronze attributed to the fourth-century BC sculptor Praxiteles and the death mask of Agamemnon. Other major collections in Greece include the Acropolis Museum (in which are preserved objects from the Acropolis site), and the Delphi Museum.

OTHER MUSEUMS

Many major museums around the world have sections devoted to ancient Greek culture. The most impressive of these include the Metropolitan Museum of Art in New York, which contains a large collection of ancient Greek vases, paintings and sculptures, and also the Hermitage Museum in Moscow, which is particularly notable for its collection of ancient Greek pottery. The Louvre in Paris, meanwhile, is home to such iconic works of art as the Venus de Milo and the Winged Victory of Samothrace, among other great sculptures. Highlights of the celebrated collections in the British Museum in London include the Parthenon Marbles (otherwise known as the Elgin Marbles), as well as sculptures, pottery and other items of great historical and artistic value.

Everyday relics

Often the most significant objects in museums are the relics of the ordinary life of ancient peoples. These vary from lengths of guttering and items of pottery (often with decorations depicting everyday life) to weapons and pieces of armour, coins, personal jewellery, eating utensils, musical instruments, children's toys, grave markers and examples of the abacus used for simple mathematical calculations.

LOST TREASURES

It is impossible to know what remains to be discovered, or what has been lost over the centuries. Some early archaeologists unwittingly destroyed much important archaeological evidence. Even Heinrich Schliemann, when he was searching for Troy, in his enthusiasm destroyed much of the evidence he was interested in before realizing his mistake. Many artefacts recovered by him subsequently disappeared during the Second World War.

THE ELGIN MARBLES

The Elgin Marbles is the name given to a collection of statues and stone reliefs that were brought from Athens to London by Lord Elgin in the early nineteenth century. Worried that the marbles risked being damaged or lost as the Turks were then using the Parthenon for target practice, Lord Elgin won permission to remove these priceless works of art, which were subsequently installed in the British Museum in London.

The Parthenon Frieze

The most important feature of the collection is the Parthenon Frieze, a magnificent series of stone reliefs that was originally placed high up on the outer walls of the Parthenon on the Acropolis, running right round all four sides of the temple just below the ceiling of the colonnade. Divided into sections, it depicts a procession of worshippers during the festival of the Great Panathenaea, which took place every four years in honour of Athena, the patron goddess of the city.

Controversy

The continued presence of the Parthenon Marbles at the British Museum in London has long been a source of controversy. The Greek government has pressed for their return, but without success.

FIND OUT MORE

BOOKS

Aegean Art and Architecture, Donald Preziosi and Louise Hitchcock, 1999

Alexander the Great, Robin Lane Fox, 2004

Ancient Greece, John Ellis Jones, 1992

Ancient Greece at a Glance, John Malam, 1998

A Visitor's Guide to Ancient Greece, Lesley Sims, 2003

Cassell Dictionary of Classical Mythology, Jenny March, 2001

Eyewitness Guides: Ancient Greece, Anita Ganeri, 1993

Greek History, Robin Osborne, 2004

Greek Life, John Guy, 1998

History of the Peloponnesian War, Thucydides, 1970

Illustrated Guide to Greek Myths and Legends, C. Evans, A. Millard and R. Matthews, 1985

The Ancient Greece of Odysseus, Peter Connolly, 1999

The Ancient World: Greece, Robert Hull, 1997

The Archaeology of Ancient Greece, James Whitley, 2001

The Cambridge Illustrated History of Ancient Greece, Paul Cartledge, 2002

The Complete World of Greek Mythology, Richard Buxton, 2004

The Greek Myths, Robert Graves, 1992

The Greeks, Susan Peach and Anne Millard, 1990

The Groovy Greeks, Terry Deary, 1995

The Iliad, Ian Strachan, 1997

The Oxford Companion to Ancient Civilization, Simon Hornblower and Antony Spawforth, 2004

The Penguin Historical Atlas of Ancient Greece, Robert Morkot, 1997

The Wars of the Ancient Greeks, Victor Davis Hanson and John Keegan, 2001

WEBSITES

The following are a brief selection of useful websites about ancient Greece:

www.bbc.co.uk/schools/ancientgreece
(general information for children)

www.crystalinks.com/greece.html
(general information)

www.culture.gr
(Greek Ministry of Culture's official site)

www.historyforkids.org
(general information for children)

www.historylink101.com/ancient_greece.htm
(general information)

www.metmuseum.org/explore/Greek/Greek1.htm
(Metropolitan Museum's collection of Greek art, with other information)

www.perseus.tufts.edu
(classic texts online)

GLOSSARY

Acanthus A plant with distinctive shaped leaves, often depicted on capitals of Greek columns.

Acropolis A fortified hill inside a city, often a sacred area.

Agora The marketplace in a Greek city.

Algebra A branch of mathematics in which numbers are represented by letters.

Amphitheatre A large semi-circular stadium in which plays and other entertainments are presented.

Amphora A two-handled jar with a narrow neck, used for holding wine and other liquids.

Andron The dining-room in a private home where the master of the house entertained male friends.

Archaic Period The period of ancient Greek history spanning the years c. 800–500 BC.

Archon A senior official of the Athenian state during the Archaic Period.

Aristocracy A state in which power is held by certain rich ruling families.

Aristocrat A rich landowner.

Aryballos A perfume pot, often fancifully shaped as an animal or mythical beast.

Assembly A gathering of Athenian citizens, where debates took place and votes were held on important matters.

Atlantes A column carved in the shape of a male figure.

Barbarian A person from outside Greece, so-called because their languages sounded like 'bar-bar' to the Greeks.

Black figure ware A style of pottery with black figures painted on a red background.

Bronze Metal made from copper, lead, tin and zinc.

Bronze Age The period of ancient Greek history spanning the years c. 2900–c. 1100 BC.

Capital The head of a pillar or column.

Caryatid A column carved in the shape of a female figure.

Cella The main room inside a temple.

Chiton A garment worn by Greek men and women, fastened at the shoulders and belted around the waist.

Chorus A group of actors who spoke, danced and sang in unison in Greek plays.

Citizen A free man over the age of eighteen, with the right to own property and take part in political and legal affairs.

City-state A Greek city and the surrounding area, politically independent of its neighbours.

Classical Period The period of ancient Greek history spanning the years c. 500–336 BC.

Colonnade A line of columns supporting a roof, wall, arch, etc.

Column A stone pillar supporting an arch or roof.

Comedy A form of drama of light or amusing character.

Corinthian An order of classical architecture, characterized by columns with capitals decorated with carved acanthus leaves.

Council A body of five hundred Athenian citizens who enacted decisions made by the Assembly.

Cuirass A body-shaped piece of armour worn by Greek soldiers to protect the chest and back.

Cult statue A statue in a temple to which prayers were addressed.

Dark Ages The period of ancient Greek history spanning the years c. 1100–800 BC.

Democracy A system of government in which leaders are voted into power by citizens.

Diadochi The generals who assumed command of areas of Alexander the Great's empire after his death.

Doric An order of classical architecture, characterized by columns with undecorated capitals.

Electrum An alloy of gold and silver used to make early Greek coins.

Ephebe A young Athenian military recruit.

Epic A long poem telling a story about gods or heroes.

Epinetron A simple semi-circular device used by Greek women when preparing wool for spinning.

Faience A form of glazed earthenware pottery produced by Minoan craftsmen and others.

Fresco A wall painting in which the paint is applied directly to wet plaster.

Frieze A band of painted or sculpted decoration placed along the upper edge of a wall.

Furies Three terrifying demi-goddesses believed to pursue and punish murderers.

Galley An oar-powered warship, usually also equipped with a sail.

Geometry A branch of mathematics relating to shapes and angles.

Grammatistes A teacher of reading, writing and mathematics.

Greaves Bronze shin guards worn in battle by Greek soldiers.

Griffin A mythical creature with the body of a lion and the head and wings of an eagle.

Gymnasium A building in which people met for physical exercise and, often, intellectual gatherings.

Gynaecum The women's quarters in a private house.

Hellene The name by which the Greeks called their own race.

Hellenistic Age The period of ancient Greek history spanning the years 323–30 BC.

Herm A small statue of the god Hermes, placed by the front door of private homes to guard the household.

Hetaerae A female slave who was hired to entertain diners with music, dance and conversation.

Himation A type of cloak worn by Greek men and women, usually draped over the left shoulder.

Hippocamp A mythical seahorse with a horse's front legs and a fish's tail.

Hoplite A foot soldier who fought with a spear and shield.

Ionic An order of classical architecture, characterized by columns with scroll capitals.

Javelin A long spear.

Kitharistes A teacher of music.

Kouros A statue of a naked

boy, with hands at his side and one leg advanced.

Krater A large pottery bowl in which wine and water were mixed.

Kylix A shallow two-handled drinking cup.

Labyrinth A maze of corridors, such as that built to house the Minotaur of Crete.

Logic The use of thinking to understand the workings of the universe.

Lyre A musical instrument resembling a small harp.

Lyric A form of poetry usually performed to lyre music.

Maenad A wild woman, a follower of Dionysus, the god of wine.

Mercenary A soldier who was paid to fight for a particular army.

Mural A wall painting in which the paint is applied to dry plaster.

Muses Nine beautiful goddesses, patrons of the arts.

Nymph A spirit of nature.

Oligarchy A state in which power is held by a select group of individuals.

Oracle A holy place where priests and priestesses were consulted for advice about the future.

Orchestra The circular central stage in a Greek theatre.

Order A style of architecture.

Ostracism The practice of voting against unpopular political figures in the hope that they might be banished.

Ostrakon A piece of pottery upon which a citizen might scratch the name of a politician and thus vote that he might be banished.

Paidogogos A domestic slave with duties to accompany the sons of a wealthy household to school.

Paidotribes A teacher of physical exercise.

Palaistra A small building used for athletics and other physical exercise.

Pankration A form of hand-to-hand combat.

Pediment The triangular end gable of a building, or the raised triangle above a doorway.

Peplos A long tunic worn by Greek women, predating the chiton.

Peristyle A row of columns around the outside of a temple.

Phalanx A military formation in which foot soldiers fought in closely-packed ranks.

Philosopher A person who engages in study of the world and its workings.

Polis The Greek word for a city-state.

Pyxis A small casket in which Greek women kept cosmetics.

Red figure ware A style of pottery in which red figures were painted on a black background.

Relief A stone carving in which the stone around the figures is chipped away.

Slave A person who was bought and sold to perform domestic or other work on the orders of his or her owner.

Soothsayer A person who claims to be able to predict the future.

Stele A stone slab marking a grave.

Stoa A long colonnaded passage in which ancient Greeks passed the time of day or did business.

Strategos A member of the group of ten Athenian generals who were elected to make decisions in time of war.

Strigil A curved blade used to scrape olive oil and dirt from the skin.

Symposium A party at which men drank and engaged in discussion or enjoyed other entertainment.

Terracotta A brownish-red pottery used in building and crafts.

Tholos A circular building with a dome, often with pillars around the outer wall.

Tragedy A form of drama tackling serious issues, often ending unhappily.

Trireme A three-decked warship with three banks of oars.

Tyrant A ruler who enjoyed complete power over his subjects.

Volute A spiral scroll decoration used to adorn capitals and pottery.

INDEX

Achilles 69, 173

Acropolis 15, 42, 107, 160, 234, 243, 245

Aeschylus 181

Agamemnon 36, 62, 243

Alexander the Great 15, 25, 27, 43, 44, 46, 47, 104, 172, 188, 223–227

Alexandria 27, 157, 199, 224

Anaxagoras 196–197

Anaximander 197

Aphrodite 55–56, 59, 60, 62, 68, 69, 70, 76, 79, 81

Apollo 51, 52, 57–58, 61, 63, 66, 70, 72, 75, 103

Archaeologists 232

Archaic Period 39–41, 113, 141, 214

Archimedes 24, 46, 197–199

Architecture 39, 134–137, 158–161, 239

Ares 54, 55, 59–60, 68, 76

Aristophanes 182

Aristotle 14, 43, 46, 133, 188, 189, 224, 238

Armies 18, 210–213

Artemis 20, 51, 52, 57, 60–61, 63, 66, 70

Asclepius 27, 57, 75, 77, 206

Athena 15, 51, 52, 57, 60, 61, 62, 63–64, 66, 70, 75, 98, 102, 107, 123

Athens 15–18, 25, 27, 39, 40, 41–45, 63, 102, 107, 110–111,

113, 114, 116–119, 123, 124, 133, 160, 188, 189, 194, 210, 211, 216, 218, 219–220, 232, 234, 243, 245

Bronze Age 32–33, 69

Centaurs 87–88

Chimera 88

Cities 26–29

City-states 14–25, 33, 39, 40, 47, 124, 210

Classical Period 15, 42–45, 113, 217

Clothing 140–141

Colonization 10, 41

Corinth 18, 19, 25, 27, 43, 46, 73, 124, 218, 221

Crete 19–20, 32, 33, 34, 95, 100, 232, 234

Cults 105–106

Cyclopes 53, 88

Dark Ages 38

Delian League 42, 116, 118, 222

Delphi 58, 72, 103, 104, 234, 243

Demeter 52, 54, 65, 72, 73, 105

Democracy 40, 43, 110, 114–115, 116

Draco 39, 114, 119

Echidna 88

Echo 83

Education 132–133

Eons 75

Ephesus 20, 203

Dionysus 51, 55, 57, 61, 63, 66–67, 70, 72, 82, 84, 107, 175

Epicurus 189
Epidaurus 27, 75, 77, 234
Eratosthenes 199
Eros 55, 56, 69, 76
Euclid 199
Euripides 43, 44, 182, 197
Family life 127–131
Farming 120–121
Fates 81, 82
Festivals 107, 144, 175–178
Food 138–139
Funerals 150–151
Furies 81
Games 145
Gods and goddesses 50–80
Golden Age 42, 44, 45, 118
Gorgons 89, 90, 99
Government 113–115, 157
Graces 81
Griffons 90
Hades 52, 54, 65, 72, 73, 85, 98
Harpies 90
Hebe 59, 68, 77
Helios 21, 69, 77, 80
Hellenistic Period 46–47, 123, 192, 240
Hephaestos 54, 55, 59, 68–69, 79
Hera 51, 52, 53, 54, 65, 68, 69, 72, 73, 79, 80, 82, 83, 84, 93
Herakles 54, 57, 60, 61, 63, 64, 66, 70, 77, 79, 88, 92, 93–96, 104
Hermes 51, 57, 61, 63, 66, 69, 70–71, 78, 98
Herodotus 43, 44, 193, 202
Hesiod 39
Hestia 51, 52, 53, 54, 65, 66, 72, 73

Hippocrates 42, 206–207
Homer 39, 156, 171–174, 224
Horai 82
Houses 134–137
Hygieia 77
Icarus 97
Ionia 38, 41, 45
Iris 78
Jason 64, 91, 98
Jewellery 143
Language 168–169, 231, 240–242
Literature 170–174, 239
Macedonians 43, 44, 45, 46, 47, 117, 151, 219, 222, 223
Maenads 82
Markets 121–122
Medicine 157, 204–207
Medusa 64, 89, 90, 98–99
Metalwork 166–167
Miletus 20–21, 187, 197
Minoans 19, 32, 33, 34–35, 38, 216, 232
Money 123
Museums 157, 243–245
Muses 82
Music 184–185
Mycenaeans 32, 33, 34, 36–37, 38, 62, 234
Narcissus 83
Nereids 78, 83
Nike 78
Nymphs 83
Odysseus 64, 74, 91, 173–174
Olympic Games 39, 147–149, 233, 238
Olympus, Mount 51, 52, 58, 68, 69, 72, 79, 92
Oracles 103–104

Orpheus 57, 86, 98
Pan 70, 78–79
Pandora 68, 79
Parthenon 42, 102, 118, 159, 160, 232, 234, 245
Peloponnesian League 25, 41
Peloponnesian War 19, 24, 25, 42, 43, 45, 151, 194, 221–222
Pericles 42, 43, 44, 102, 118, 151, 197
Perseus 64, 89, 90, 98–99
Persians 42, 43, 44, 104, 116, 118, 119, 193, 218, 219–220, 223
Philosophers 156, 157, 186–192, 238
Plato 43, 133, 188, 189–190, 192, 203, 238
Playwrights 181–183
Politicians, Athenian 116–119
Poseidon 51, 52, 54, 65, 72, 73–74, 80, 85, 126
Pottery 164–165
Priapus 55, 79
Prometheus 79
Proteus 80
Pythagoras 40, 202, 203
Rhodes 21, 46
Romans 19, 27, 46, 47, 192, 198, 238, 239
Satyrs 84, 175
Science 157, 195–203, 239
Sculpture 162–163
Selene 80
Ships 18, 126, 216–217
Sirens 91

Slavery 111–112
Socrates 43, 133, 190, 191–192, 194, 238
Sophocles 43, 44, 182, 183
Sparta 21–24, 25, 39, 41, 42, 43, 45, 52, 60, 113, 127, 129, 131, 146, 194, 210, 218, 221–222
Sphynx 91
Sport 146–149
Stoics 192
Syracuse 24, 43, 198
Temples 101–102, 158
Thales 40, 187, 197, 203
Theatre 175–183
Themis 80, 81
Theocritus 24
Theseus 73, 95, 100
Thucydides 194
Timaeus 194
Titans 51, 52, 75, 79, 81, 88, 90
Trade 124, 216
Transport 125–126
Triton 80
Trojan War 32, 36, 56, 171, 173–174, 219
Troy 32, 62, 64, 74, 215, 232, 234–237
Underworld 65, 71, 81, 85–86, 88, 96, 150
Warfare 209–227
Wine 139
Xenophon 194
Zeus 51, 52–53, 54, 55, 57, 59, 61, 63, 65, 66, 68, 69, 70, 72, 73, 75, 78, 79, 80, 81, 83, 85, 88, 90, 92, 93, 102, 104

Look out for further titles in the Collins Gem series.

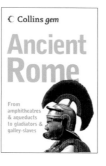

Collins *gem*

Ancient Rome

From amphitheatres & aqueducts to gladiators & galley-slaves

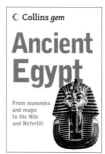

Collins *gem*

Ancient Egypt

From mummies and magic to the Nile and Nefertiti

Collins *gem*

Pirates

From corsairs and cutlasses to parrots and planks

Collins *gem*

Royal Britain

Fascinating insight into British royal life